The HISTORY of the DECLINE AND FALL of the RATERPILLAR TRACTOR COMPANY

A MODERN BUSINESS SAGA

By
MICHAEL ZINMAN

HAYDN FOUNDATION
Ardsley, New York

Library of Congress Catalog Card No.
ISBN: 0-937704-10-2

Revised second printing 2003

Manufactured in the united States of America

To
Lee L. Morgan
And
E.C. "Chappy" Chapman

Two men whose single-mindedness and imperial sense of being have made Caterpillar Tractor Co. the company it is today.

And for
Albert E. Norman

Preface to the Second Edition

Authors like to see their names in print, and the result is a very healthy and profitable Vanity Press industry.

Raterpillar seems to have been an exception to the rule. It started out as a set of xeroxed pages sent to a few friends, who passed them along to others, and I started receiving requests for additional copies. I tired of filling these requests, and looked to put the book in print.

One of the investment banking companies got hold of the xeroxes, and asked if they could use the book as a Christmas gift to their clients. They underwrote the costs of the original edition, which sold out.

Even after being away from the construction industry for some fifteen years, I still get one or two calls each month for a copy, of which I have not a one. I have finally succumbed and decided to print up another batch, to slake the thirst of my ravenous public.

I borrowed a copy of the book to use for the new edition, and read it through anew. It's still fresh as a daisy, even though the players are not. Rennie Atterbury retired from Caterpillar, as did Lee Morgan and E.C. Chapman. Al Norman is living in Boca Raton, Florida and, as far as I can learn, the other players are still around.

Not the rats, though.

Ardsley, New York
July, 2003

PREFACE

By way of explanation and apology, I would like to make known to the reader the origin of this book. It grew haphazardly. A number of the participants wanted some memorial of the events that had transpired, and this was provided by copying the documents found herein. As the demand for the story increased, the narrative you will read was added. All of the readers, however, had either knowledge of the companies or the individuals involved, and as such a seemingly awkward familiarity might be inferred in reading the narrative at this time.

I was urged to have a professional rewrite it (the narrative) but then figured it might as well be read warts and all.

M.Z.

Ardsley, New York

October 1986

Every word of the story that unfolds before you, boys and girls, is true. It has all the ingredients of a great business story: greed, cunning, global strategies, and, best of all, lawyers.

It begins and ends, not surprisingly, perhaps, with a lawyer's letter. To understand and follow all the threads and nuances that pepper this narrative, a little history is essential.

Back in 1976, Caterpillar Tractor Co. brought an action in the Federal District Court of the Southern District of New York against Earthworm Tractor Co., a small, privately held equipment trading company. I founded Earthworm in 1968 and was sole owner at that time. Caterpillar charged that Earthworm was unfairly competing with it and looked to have me disgorge all the profits Earthworm had ever made and pay them over to Caterpillar—this to redress my sins and assuage their injured sense of importance.

Caterpillar's basic accusation of unfair competition referred to a stylized "E" we were using on our letterhead which they felt looked too much like a stylized "C" used in their letterhead. I was very upset at this charge of unfair competition. I called my lawyers. I then brought a counter-suit against Caterpillar accusing them of violating the Sherman-Clayton and Robinson-Patman Acts. After five or six years in the courts, the matter was settled in a reasonably amicable way—or so I thought—after Robert Thornton, Esq., a splendid lawyer formerly associated with the firm of Nixon Mudge Rose Guthrie and Alexander in New York, assumed the position of General Counsel to Caterpillar Tractor Co. This was in 1983.

Some months after this settlement was effected, a meeting was held at my lawyer's office. A chance remark deemed, I assumed by all, to be no more or less lunatic than so many others made by the author was acted upon. It was, as you might have guessed: "Why don't we incorporate 'Raterpillar Tractor Company?'"

More months passed. And then on that fateful day, February 22, 1984, events were set in motion that, I believe, added a small measure of mirth to the world.

I put before you the unexpurgated Raterpillar story.

LAW OFFICES OF

DECHERT PRICE & RHOADS

477 MADISON AVENUE

NEW YORK, NY 10022

TELEX 64 5324 • BARDEP

(212) 308-4400

3400 CENTRE SQUARE WEST
1500 MARKET STREET
PHILADELPHIA, PA 19102
(215) 972-3400

38 SQUARE DE MEEUS, STE 1
1040 BRUSSELS, BELGIUM
(02) 511 60 40

52 BEDFORD SQUARE
LONDON WC1B 3EX, ENGLAND
01. 631. 3383

1730 PENNSYLVANIA AVENUE, N.W.
WASHINGTON, DC 20006
(202) 783-0200

800 NORTH THIRD STREET
HARRISBURG, PA 17102
(717) 233-7947

ROBERT A. COHEN

February 22, 1984

Mr. Michael Zinman
Earthworm Tractor Co. (Inc.)
495 Ashford Avenue
Ardsley, New York 10502

Re: Raterpillar Tractor Co., Inc.

Dear Michael:

Here is a copy of your Certificate of Incorporation. Be careful!

Best regards.

Sincerely,

Robert

Robert A. Cohen

RAC:mc
Encl.

Certificate of Incorporation
of
Raterpillar Tractor Co., Inc.

(Under Section 402 of the Business Corporation Law)

The undersigned, being a natural person of at least 18 years of age, and acting as the incorporator of the corporation, hereby being formed under the Business Corporation Law of the State of New York, certifies:

<u>First:</u> The name of the corporation shall be Raterpillar Tractor Co., Inc.

B062753

State of New York } ss.:
Department of State }

4726

I hereby certify that I have compared the annexed copy with the original document filed by the Department of State and that the same is a correct transcript of said original.

Witness my hand and seal of the Department of State on JAN 26 1984

Secretary of State

380807-004(12/82)

When Bob Cohen sent me the copy of the certificate of the Raterpillar incorporation, I was astounded. I had forgotten the entire matter and never expected my remarks, made the previous Fall, to have been remembered by anyone else either—to say nothing of receiving action based thereon.

This letter and its accompanying documents were irresistible. I promptly produced a Raterpillar letterhead cobbled out of a Caterpillar letterhead, xeroxed some blanks, and sent a letter off to Bob Thornton over at Cat.

RATERPILLAR TRACTOR CO.

495 Ashford Avenue
Ardsley, New York 10502
Telephone (914) 693-0400

March 1, 1984

Robert Thornton, Esq.
Caterpillar Tractor Company
100 N.E. Adams Street
Peoria, IL 61629

Dear Robert:

And you thought things would be quieter in Peoria.

I am pleased to announce the formation of a new company, duly organized to do business in the State of New York. I am somewhat unsure as to its ultimate aims, goals and aspirations, but do not fret, they will come to me. However, I assure you it will not be in the area of the manufacture, sale, or brokerage of heavy construction equipment or other enlightened Caterpillar endeavors.

As you will note, we have developed a proposed letterhead and we are considering using a disclaimer as shown below, so that there is no confusion in the minds of our respective customers.

If you can suggest any other disclaimer you wish us to make as to more clearly segregate the respective identities of our two companies, I will be disposed to consider any such request in a conscientious and forthright manner.

With kindest regards and best wishes for your continued success, I am and shall remain (for the present at least),

Very truly yours,

RATERPILLAR TRACTOR CO.

Michael Zinman,
President

MZ:llc

A Modest Disclaimer: Raterpillar Tractor Co. is in no way connected with Caterpillar Tractor Co. Caterpillar Tractor Co. sells "Cats" — Raterpillar Tractor Co. sells rats.

Cohen followed up his letter with some necessary tax forms he had received in the normal course of events from the State of New York.

This is a *complete* corporate history.

LAW OFFICES OF

DECHERT PRICE & RHOADS

477 MADISON AVENUE
NEW YORK, NY 10022
TELEX 64 5324 • BARDEP
(212) 308-4400

3400 CENTRE SQUARE WEST
1500 MARKET STREET
PHILADELPHIA, PA 19102
(215) 972-3400

38 SQUARE DE MEEUS, STE 1
1040 BRUSSELS, BELGIUM
(02) 511 60 40

52 BEDFORD SQUARE
LONDON WC1B 3EX, ENGLAND
01. 631. 3383

1730 PENNSYLVANIA AVENUE, N.W.
WASHINGTON, DC 20006
(202) 783-0200

800 NORTH THIRD STREET
HARRISBURG, PA 17102
(717) 233-7947

ROBERT A. COHEN

March 6, 1984

Mr. Michael Zinman
Earthworm Tractor Co. (Inc.)
495 Ashford Avenue
Ardsley, New York 10502

Dear Michael:

Enclosed is some "Corporation Tax Account Information" from the New York State Department of Taxation and Finance with regard to Raterpillar Tractor Co., Inc. Please let me know what address I should indicate for the corporation. We will apply for a federal employer ID number. I assume you don't care about the accounting period, at least for now.

Best regards.

Sincerely,

Robert A. Cohen

RAC:mc
Encl.

New York State Department of
TAXATION and FINANCE
Processing Division - Corporation Tax
State Campus, Albany, New York 12227

Corporation Tax Account Information
If corrections in our records are necessary return this form as soon as possible to ensure proper identification.

• RATERPILLAR TRACTOR CO., INC.
DECHERT ETAL
477 MADISON AVE
NEW YORK NY 10022

Identification Number	File Number	Filing Period
CT-1556754	AA06	12-31-84

1. Correct Address

2 Federal or Corrected Employer ID Number

3. ☐ - EIN not applicable

4 Correct/New Filing Period

The above information is currently being used by N.Y.S. Corporation Tax to identify your account. To insure proper credit to your account the above identification and file number should be included on all future reports and correspondence until advised otherwise. Use this form and the enclosed return envelope to notify N.Y.S. Corporation Tax of any incorrect or revised information. See below.

- Box 1 - If your computer printed mailing address or ZIP code above is wrong, please make corrections in this box.
- Box 2 - Enter your federal employer identification number. If you have applied for a federal EIN but not yet received one, keep this form and return it when the information is available.
- Box 3 - Check this box if you do not intend to apply for a federal employer ID number.
- Box 4 - If the computer printed filing period is incorrect enter the correct/new period in this box. For additional information see the back of this form.

Corporation Tax Section

If you are located in	For forms and publications call	For information call
New York State	(212) 488-3608	(212) 488-3400
Outside New York State	1-800-462-8100 TOLL FREE	1-800-342-3536 TOLL FREE
	(518) 457-0111	(518) 457-0111

Enclosure: Return Envelope

CT-198 (7/83)
9

Caterpillar reacts. Al Robin called Bob Cohen on April 3rd and told him that Caterpillar had sent my letter to Thornton to him and Robin wanted to bring an action against me.

Robin had been one of the lawyers that Caterpillar used in the *Earthworm* case [77 Civ 794 (LPG) for those hardy souls who want to know everything there is to know], but since he handled only the trademark aspects of the matter, Robin had not had all that much to do with the final negotiations.

Robin had a very sanguine opinion of the matter and was of the view that he could wipe the floor with me in any litigation. Unfortunately, Cohen did not disabuse me of this point but did mention that Caterpillar management had told Robin they "did not want Zinman digging his heels in again" (presumably for another protracted lawsuit) and asked if I would change the name of the company, the joke having been made.

Robin followed up the call with a letter dated April 6, 1984.

LAW OFFICES OF
ALBERT ROBIN
1270 AVENUE OF THE AMERICAS
SUITE 220
NEW YORK, NEW YORK 10020
(212) 977-8844

Albert Robin
Howard B. Barnaby

TELEX
126337

CABLE
RITEROBIN
NEW YORK

April 6, 1984

Robert A. Cohen, Esq.
Dechert, Price & Rhoads
477 Madison Avenue
New York, New York 10022

Re: Caterpillar Tractor Co. v.
Raterpillar Tractor Co., Inc.
Our File C450-000

Dear Bob:

On Tuesday, April 3, 1984, I telephoned you to advise you of Caterpillar Tractor Co.'s position with respect to Michael Zinman's recent incorporation of Raterpillar Tractor Co., Inc. You promised to look into the matter and get back to me. Since I have not heard from you, I am writing to confirm what I told you then.

As I mentioned on Tuesday, it is possible that Michael Zinman does not seriously intend to use the "Raterpillar" name. In this connection, I thoroughly enjoyed Michael's letter to Robert Thornton of March 1, 1984. If this is the case, then there should be no problem in changing the name to one which does not include "Raterpillar" or any other simulation of Caterpillar.

As I also explained on Tuesday, if the "Raterpillar" name is not changed, my client would view use of that name as a violation of its rights and take appropriate action to obtain cessation of such use.

In my view, any use by your client of "Raterpillar" could not help but constitute infringement and unfair competition and/or injury to business reputation and dilution. I hope I will have your prompt assurances that your client will change the name of his company to eliminate "Raterpillar" therefrom.

Sincerely yours,

Al

Albert Robin

AR/cs

cc: Eugene C. Goodale, Esq.

LAW OFFICES OF

DECHERT PRICE & RHOADS

477 MADISON AVENUE

NEW YORK, NY 10022

TELEX 64 5324 - BARDEP

(212) 308-4400

3400 CENTRE SQUARE WEST
1500 MARKET STREET
PHILADELPHIA, PA 19102
(215) 972-3400

36 SQUARE DE MEEUS, BTE 1
1040 BRUSSELS, BELGIUM
(02) 511 60 40

52 BEDFORD SQUARE
LONDON WC1B 3EX, ENGLAND
01. 631. 3383

1730 PENNSYLVANIA AVENUE, N.W.
WASHINGTON, DC 20006
(202) 783-0200

800 NORTH THIRD STREET
HARRISBURG, PA 17102
(717) 233-7647

ROBERT A. COHEN

April 9, 1984

Albert Robin, Esq.
1270 Avenue of the Americas
Suite 220
New York, New York 10020

Re: Caterpillar Tractor Co. v.
Raterpillar Tractor Co., Inc.
Your File C450-000

Dear Al:

Your letter of April 6th accurately reflects our telephone conversation of April 3rd. I should be able to get back to you by the end of this week.

Best regards.

Sincerely,

Robert

Robert A. Cohen

RAC:mc

cc: Michael Zinman
w/encl.

The plot thickens. I knew, of course, that I could really do nothing with Raterpillar, but much enjoyment could be (and so often is) obtained from doing nothing. I also knew that it was impractical for me to think I could do nothing. I must now admit (modest though I am) that I conceived the idea of selling Raterpillar to Caterpillar at this time with the sole motive being the placement of a tombstone announcement in *The Wall Street Journal* as a final gesture to the episode.

On April 19, 1984, a letter was sent to Bob Thornton. I also asked Bob Cohen to research my position, and his comments follow.

RATERPILLAR TRACTOR CO.

495 Ashford Avenue
Ardsley, New York 10502
Telephone (914) 693-0400

April 19, 1984

Robert Thornton, Esq.
Caterpillar Tractor Company
100 N.E. Adams Street
Peoria, IL 61629

Dear Bob:

I received copies of Al Robin's correspondence with Bob Cohen (I hope he partook of the wisdom of Walt Robinson's good counsel) and have given the entire Raterpillar matter careful and considered thought.

First, although some fertile ideas have admittedly been kicked around by the Raterpillar brain trust, other than printing our stationery and designing some advertising layouts for the product (rat of the week, adopt a rat, etc.) we have done little else with the Company.

I take exception to Al's position that Caterpillar could make Raterpillar bite the poisoned cheese. I think that the pleasure of living (read the ability to laugh at yourself and others - a commodity in precious short supply in Peoria) can not be denied. I sincerely feel the entire matter is in good, relatively clean (rats are dirty) fun, and I am certain that Raterpillar could be run profitably, particularly with the future possibilities of tee shirts and fur products made from rat skins.

I am not particularly interested in changing the name of the Company as you suggest. Also, you must realize that by my changing the name, such an action in no way protects Caterpillar from someone else adopting it (as a matter of fact I have severl other parties interested in the Company waiting breathlessly in the wings; but have no fear I told them that Caterpillar has first choice in this matter).

I would be willing to sell Raterpillar to you. I have checked and found that costs incurred by me to date amount to $383.80 which is comprised of $300 for incorporation, $83 for stationery, and 80¢ for postage. These costs will of necessity increase as I must add postage to both you and Al for this letter. Therefore, I offer Raterpillar to you at cost $384.20. While I would think it an egregiously cruel act on your part to haggle, I do wish you to know that I would not dismiss a counter offer out of hand.

A Modest Disclaimer: Raterpillar Tractor Co. is in no way connected with Caterpillar Tractor Co. Caterpillar Tractor Co. sells "Cats" — Raterpillar Tractor Co. sells rats.

RATERPILLAR TRACTOR CO.

495 Ashford Avenue
Ardsley, New York 10502
Telephone (914) 693-0400

Caterpillar Tractor Company -2- April 19, 1984

Lastly I am comforted and will always be comforted in the cold winter months ahead by conjuring up in my mind the picture of you going before the board of Caterpillar (should you be inclined to do so) to ask for the necessary funds to bring this event to a brilliant conclusion.

Cordially,

RATERPILLAR TRACTOR CO.

Michael Zinman,
President

MZ/lt

cc: Albert Robin, Esq.
Robert Cohen, Esq.

A Modest Disclaimer: Raterpillar Tractor Co. is in no way connected with Caterpillar Tractor Co. Caterpillar Tractor Co. sells "Cats" — Raterpillar Tractor Co. sells rats.

3400 CENTRE SQUARE WEST
1500 MARKET STREET
PHILADELPHIA, PA 19102
(215) 972-3400

38 SQUARE DE MEEUS, BTE 1
1040 BRUSSELS, BELGIUM
(02) 511 80 40

32 BEDFORD SQUARE
LONDON WC1B 3EX, ENGLAND
01. 631. 3383

LAW OFFICES OF
DECHERT PRICE & RHOADS
477 MADISON AVENUE
NEW YORK, NY 10022
TELEX 64 5324 • BARDEP
(212) 308-4400

1730 PENNSYLVANIA AVENUE, N.W.
WASHINGTON, DC 20006
(202) 783-0200

800 NORTH THIRD STREET
HARRISBURG, PA 17102
(717) 233-7947

ROBERT A. COHEN

April 19, 1984

Federal Express

Mr. Michael Zinman
Earthworm Tractor Co.
495 Ashford Avenue
Ardsley, New York 10502

Dear Michael:

You've asked me to consider the legal implications of using the "Raterpillar Tractor Co." name to advertise the sale of rats to purchasers of tractors (either in trade journals or in individual mailings.) The name Raterpillar, which we can agree is a take-off on the name Caterpillar Tractor Co., and the proposed scheme, are amusing to us and would no doubt be amusing to tractor purchasers who are familiar with Caterpillar. A crucial element in the discussion which follows is the notion that those reached by the Raterpillar sales pitch for rats will know that you are behind it, and that that will, in some perhaps indirect, subliminal way, result in the sale by you of more tractors.

Counsel for Caterpillar has stated that Caterpillar will sue to prevent use of the name on theories of injury to business reputation and dilution and has stated that he "can't lose."

I have had research done to determine if, indeed, Caterpillar would be able effectively to state a cause of action should you proceed with the Raterpillar scheme. The research shows that under the theories of suit that Caterpillar's counsel cited to me, there is a strong liklihood that a court would find the use of the name Raterpillar as proposed to be illegal.

Mr. Michael Zinman
April 19, 1984
Page Two

One theory of recovery that is available to Caterpillar is provided by the federal Lanham Act. That Act is aimed at preventing use of a trade name that "infringes," in that the name tends to mislead, deceive or confuse the public as to the source of a product's or service's origin, sponsorship or association. Courts are persuaded that confusion is likely where they find one or more of the following factors: a very distinctive and recognizable tradename belonging to the plaintiff; close similarity of the plaintiff's and defendant's names; similarity of products or services associated with the names; and, intent of the defendant to take advantage of the plaintiff's well recognized name and reputation.

New York law provides a cause of action for infringement under section 368-d of the New York General Business Law, entitled "Injury to Business Reputation and Dilution." The courts, in applying that section, have interpreted it to require a showing by the plaintiff that its tradename is very distinctive and that it is likely that the defendant's name is "diluting" the commercial value of plaintiff's name. A cause of action is easier to make out under this section than under federal law because the meaning of the term dilution is not well defined or rigidly applied. For example, courts have defined "dilution" as the "whittling down" of the identity of the name, the "blurring" of the name's product identification or the "tarnishing" of the affirmative associations that the name conveys. Such attempts to define dilution are not very helpful in giving the term meaningful content; however, from the courts' treatment of the term it is clear that dilution is a subtle form of infringement that does not require any proof that confusion of the public is likely.

I believe that Caterpillar's chances of success are greater under New York law than federal law because it would be difficult to prove, as required under federal law, that it is likely that tractor dealers would confuse Raterpillar with Caterpillar as the sponsor of the advertisement for the sale of rats. To determine Caterpillar's likelihood of success under New York law, it is helpful to look at cases that have similarities to our situation. There are two dilution cases that involve the defendant's use of a name that parodies, puns or satirizes a well known trade name. They are different from our situation in that, although aimed at commercial gain from the play on the tradename, neither involved the selling of competing products (even indirectly).

Mr. Michael Zinman
April 19, 1984
Page Three

For example, the Court in Tetley Tea Bags v. Topps Chewing Gum considered whether the defendant, a chewing gum producer, diluted the name Tetley Tea Bag by including with its product, gum, stickers resembling the Tetley box with the name "Petley Flea Bags". The court decided there was no dilution because the blatant humor prevented blurring of the name and because the defendant was not selling a similar product in the plaintiff's market. Also, the court in Girl Scouts of America v. Personality Posters, Inc. found that the sale of posters showing a pregnant Girl Scout, with the scout motto "be prepared" was not dilutive of the Girl Scout name. The court allowed this satire as a protected form of free expression. It is important to note that in this case, also, there was no competition between the plaintiff and defendant.

These two decisions do not mean, however, that courts always find that parodies of tradenames done in jest do not constitute infringement. Where the court finds that the defendant acted in bad faith, by intending that the public believe that the plaintiff was associated with the parody or satirical depiction, infringement of the plaintiff's tradename is likely to be found, despite the element of parody or satire.

Cases where a court found that a name did dilute the plaintiff's name further reveal the factors on which courts focus to find the likelihood of dilution. For example, the court concluded that the name Kids 'R' Us diluted the plaintiff's name Toys 'R' Us because (1) the plaintiff's name was very distinctive, 2) the defendant's name, though distinguishable, made use of the distinctive, commercially valuable feature of the name, 3) the products sold by both were similar and 4) the defendant knew of the plaintiff's, name and intended to profit from it. Likewise, the court decided that the name Nina of California diluted Nina Needlepoint Inc. because the diluting name was very similar to the plaintiff's, it was chosen for its similarity to the plaintiff's name and the products sold under both names were the same. At least three other cases that I have at hand indicate that the same factors are crucial: similarity of name, similarity of product and intent of the defendant to profit from the plaintiff's name.

Applying these three factors to our situation, it appears that there is a substantial risk that the name Raterpillar would be considered dilutive and an infringement

Mr. Michael Zinman
April 19, 1984
Page Four

on the Caterpillar name. That conclusion follows because the names are nearly identical, the same products, tractors, are (or it is hoped will be) associated with both names (albeit indirectly associated with Raterpillar), and a bad faith intent to profit from the Caterpillar name could be inferred.

Although it appears likely that the Raterpillar advertisement would constitute tradename infringement, there is still the question of the kind of award Caterpillar could obtain. If the risk were only that an injunction would issue, there would be little to lose by going ahead with the plan. However, both the Lanham Act and New York law allow punitive damages. An injunction is by far the most common remedy; cases show an award of punitive damages only for egregious violations constituting fraud, such as where the defendant counterfeited the labels of another product. However, the court could find that, since this is the second time you would have been involved in an alleged attempt to appropriate the business of Caterpillar by infringing on its tradename, (the first event resulting in a preliminary injunction followed by a stipulated consent to cease the objectionable conduct), punitive damages would be appropriate. This is, of course, speculation, since no case has been found that presents an instance of a repeat offender.

The clear conclusions that I am able to reach are that the scheme is an invitation to litigation and that in the suit there would be some sound arguments on both sides. Given the past conflicts with Caterpillar on the issue of trademark infringement, it is unlikely it would be willing to settle. The complicating factor in deciding whether to go ahead with this scheme is the threat of punitive damages. The degree of probability that the court would award them is too speculative to hazard a prediction; nonetheless, the fact remains that they are available.

Mr. Michael Zinman
April 19, 1984
Page Five

Please call me to discuss this matter. I have told Caterpillar's counsel that I would inform him by the end of the week as to whether you will continue to use the Raterpillar name.

Best regards.

Sincerely,

Robert

Robert A. Cohen

William Miller. William Miller is a marvelous man. He teaches school in the South Bronx [depicted with clarity and accuracy in the movie *Fort Apache, the Bronx* for those readers who need an added point of reference]. Mr. Miller is also a friend of mine and visits from time to time. He saw some of the previous correspondence. He liked it, and, like the NRA, wanted to do his share.

From the genius of Miller's mind came the Raterpillar logo. Also with it came a dozen or so Raterpillar T-shirts, with the logo emblazoned—black on yellow. Wearing them became a badge of honor.

A few were sent into the world.

CATERPILLAR SELLS "CATS"
RATERPILLAR
RATERPILLAR SELLS "RATS"

RATERPILLAR TRACTOR CO.

495 Ashford Avenue
Ardsley, New York 10502
Telephone (914) 693-0400

June 11, 1984

Albert Robin, Esq.
Suite 1220
1270 Avenue of the Americas
New York, NY 10020

Dear Al:

Stick with me and you'll never go naked.

Cordially,

RATERPILLAR TRACTOR CO., INC.

Michael Zinman,
President

MZ:llc
Enclosure

A Modest Disclaimer: Raterpillar Tractor Co. is in no way connected with Caterpillar Tractor Co. Caterpillar Tractor Co. sells "Cats" — Raterpillar Tractor Co. sells rats.

RATERPILLAR TRACTOR CO.

495 Ashford Avenue
Ardsley, New York 10502
Telephone (914) 693-0400

June 11, 1984

Robert Thornton, Esq.
Caterpillar Tractor Company
100 N.E. Adams Street
Peoria, IL 61629

Dear Bob:

Just a sweet thought to brighten up your day.

Cordially,

RATERPILLAR TRACTOR CO., INC.

Michael Zinman,
President

MZ:llc
Enclosure

A Modest Disclaimer: Raterpillar Tractor Co. is in no way connected with Caterpillar Tractor Co. Caterpillar Tractor Co. sells "Cats" — Raterpillar Tractor Co. sells rats.

RATERPILLAR TRACTOR CO.

495 Ashford Avenue
Ardsley, New York 10502
Telephone (914) 693-0400

June 11, 1984

Robert Cohen, Esq.
Dechert, Price and Rhoads
477 Madison Avenue
New York, NY 10022

Dear Bob:

Some things need no further amplification.

RATERPILLAR TRACTOR CO., INC.

Michael Zinman,
President

MZ:llc
Enclosure

A Modest Disclaimer: Raterpillar Tractor Co. is in no way connected with Caterpillar Tractor Co. Caterpillar Tractor Co. sells "Cats" — Raterpillar Tractor Co. sells rats.

From my association with Caterpillar, I am certain that what follows is the only genuinely funny letter ever written by them. I still don't believe it. Eugene Goodale, Esq., the writer, was in fact the lawyer at Caterpillar who originally started the action in the mid-seventies on the Earthworm "E" when he interposed an objection to the issuance of a Service Mark by the Patent Office that we had applied for.

The plaque Goodale referred to was one I had made after the settlement of the *Earthworm* case, which had on it the citation of the case and a small ceramic earthworm mounted. The worm was fired in yellow glaze. It was in quite good taste, I thought.

I do not know if Mr. Goodale is still employed by Caterpillar.

CATERPILLAR TRACTOR CO.

Peoria, Illinois 61629

July 9, 1984

Mr. Michael Zinman
495 Ashford Avenue
Ardsley, New York 10502

Re: Raterpillar Tractor Co.
Our File: 540.372

Dear Mr. Zinman:

Bob Thorton has asked that I respond, even if belatedly, to your offer of April 19, 1984 to sell your newly-initiated venture into the rat field. On behalf of both Bob and Al Robin, I can report that the tee shirts were received and have taken their places in the respective trophy cases.

Off the record, I believe the shirts received much the same reaction as the retirement shirt you sent to Kenny Ames many years ago. As you mentioned to me some years ago, many large companies have lost a sense of humor. The Raterpillar tee shirt reaction ranged from laughter to !#¢&*@$!!. The wooden placque you sent to Al as a memento of the trademark case was priceless - of course, varied reactions came forth. With great reluctance, Al presented the placque to Les Barra at the time of his retirement from the Trademark Section and Caterpillar. I feel like the perpetual "bridesmaid" relative to your creative(?) and ingenious(?) mementos - close, but never at the altar.

Be that as it may, back on the record, as regarding the outstanding offer which is not quite binding --- after seven straight quarters of no profit, I would venture to say that your sales and profit picture may be better than Caterpillar's. Maybe you should pay us to take this company off your hands - no sense pouring good money down a rat hole.

While your offer may be considered generous by some and outrageous by others, I propose a counteroffer in view of the economic times and past seven quarters. I suggest that we split the offer 50-50 (after adding a modest price for tee shirts) such that for the sum of two hundred dollars U.S. ($200.00) you deed Raterpillar to Caterpillar. My estimate for tee shirts may be off, but let's not quibble over a rat's whisker. After all, this is not a cat and mouse game, as some might infer.

CATERPILLAR TRACTOR CO.

Mr. Michael Zinman -2- July 9, 1984

I believe that this offer is equitable, if not generous. Both parties will have expended a modest fee in this exercise of corporate mirth.

I look forward to your response to my counteroffer. However, if a response is not immediately forthcoming, you can rest assured that I have not entered an early follow-up to this letter. The fact is, my follow-up may get lost in the maze between enter and exit.

Yours for a brighter future.

Very truly yours,

Eugene C. Goodale
Supervising Patent Attorney
Trademarks and Copyrights

ECG:kgb
Telephone: (309) 675-5089

cc: R. R. Thornton - AB7C
A Robin - New York

When I received Goodale's letter, I knew the springe had been sprung. As they say in France, I had 'em by the *baytzim*. However, Earthworm was in the throes of trying to bring out an initial public issue of its shares, and the Raterpillar matter had to be caged until the Fall.

RATERPILLAR TRACTOR CO.

495 Ashford Avenue
Ardsley, New York 10502
Telephone (914) 693-0400

October 10, 1984

Mr. Eugene Goodale
Supervising Patent Attorney
Caterpillar Tractor Company
100 N.E. Adams Street
Peoria, ILL 61629

Dear Eugene:

How I have pondered your letter of July 9, 1984, and how humble I am before it! I had hoped to respond sooner, but great enterprises have stood in the way. You will note, I missed the opportunity to rebut the argument of seven straight quarters of Caterpillar's losses (you reported a profit for the second quarter) but now, alas, business has become once again tough and unprofitable.

In any event, we should put this child, Raterpillar, to bed, and as not to unduly prolong this event, I accept your offer of $200. However, I believe certain formal steps are necessary, inasmuch as shares must be transferred and representations made. And also, there is the matter of the stationery. While this will become yours, who pays to pack it up and ship it to Peoria? Or, will you send one of the corporate jets to pick up the material (there is an airport nearby)?

My eminent counsel has suggested that you prepare a suitable document with all necessary claims, covenants, and mutual releases you would like me to agree to (having to do with Raterpillar -- don't get too excited), and submit it to me for execution.

I would welcome the opportunity of inviting you to New York for the closing, so that this marvelous event can be duly celebrated (perhaps a luncheon at McDonald's). However, I guess I could live with signing the documents and sending them back in the mail.

I do look forward to hearing from you in this matter and have sent you, under separate cover, a small memento of the great firm, Earthworm Tractor Company.

Cordially,

RATERPILLAR TRACTOR CO.

Michael Zinman
President

MZ:llc

By now news of the Raterpillar affair had reached the trade, and there were many requests for letterheads. Bob Cleveland is the president of a Caterpillar dealership and asked if I would send him examples of what we were using.

RATERPILLAR TRACTOR CO.

495 Ashford Avenue
Ardsley, New York 10502
Telephone (914) 693-0400

November 29, 1984

Mr. Robert Cleveland, Jr.
President
H.O. Penn Machinery Company
Interstate 684 at Route 22
Armonk, N.Y. 10504

Dear Bob:

This is the letterhead we use in dealing with the great unwashed, to avoid confusion.

Very truly yours,

RATERPILLAR TRACTOR CO.

Michael Zinman,
President

MZ:llc

A Modest Disclaimer: Raterpillar Tractor Co. is in no way connected with Caterpillar Tractor Co. Caterpillar Tractor Co. sells "Cats" — Raterpillar Tractor Co. sells rats.

RATERPILLAR TRACTOR CO.

495 Ashford Avenue
Ardsley, New York 10502
Telephone (914) 693-0400

November 29, 1984

Mr. Robert Cleveland, Jr.
President
H.O. Penn Machinery Company
Interstate 684 at Route 22
Armonk, New York 10504

Dear Bob:

Wanna buy a rat?

Very truly yours,

RATERPILLAR TRACTOR CO.

Michael Zinman,
President

MZ:llc

More tax matters. . .

LAW OFFICES OF

DECHERT PRICE & RHOADS

477 MADISON AVENUE
NEW YORK, NY 10022
TELEX 64 5324 • BARDEP

(212) 308-4400

3400 CENTRE SQUARE WEST
1500 MARKET STREET
PHILADELPHIA, PA 19102
(215) 972-3400

38 SQUARE DE MEEUS, STE 1
1040 BRUSSELS, BELGIUM
(02) 511 80 40

32 BEDFORD SQUARE
LONDON WC1B 3EX, ENGLAND
01. 631. 3383

1730 PENNSYLVANIA AVENUE, N.W.
WASHINGTON, DC 20006
(202) 783-0200

800 NORTH THIRD STREET
HARRISBURG, PA 17102
(717) 233-7947

ROBERT A. COHEN

December 3, 1984

Mr. Michael Zinman
Earthworm Tractor Co. Inc.
495 Ashford Avenue
P.O. Box 104
Ardsley, New York 10502

Re: Raterpillar Tractor Co., Inc.

Dear Michael:

Enclosed is a communication from the State of New York Department of Taxation and Finance. Until you sell the stock to Caterpillar, you will continue to get such missives.

Best regards.

Sincerely,

Robert A. Cohen

RAC:bf
Enclosure

New York State Department of
TAXATION and FINANCE
Common Business ID Project
State Campus, Albany, N.Y. 12227
Telephone: (518) 457-2040

6008789948

RATERPILLAR TRACTOR CO., INC.
*DECHERT ETAL
477 MADISON AVE
NEW YORK NY 10022

The Tax Department is in the process of updating account information and standardizing identification of accounts for all business taxpayers.

The account information shown below represents the identification of at least one of your tax accounts. If you have accounts under other names or identification numbers, you will also be receiving notices for those accounts. If you receive more than one notice, complete one, attach all others, and return to this office.

Please check the account information shown below to verify its accuracy

Name: RATERPILLAR TRACTOR CO., INC.

Number(s)	Tax Type	Number(s)	Tax Type
CT-1556754 AA06	CORPORATION		

*(Asterisk) indicates inactive account. If account should be reactivated, place an "X" in the box to the right of asterisk.

If the information shown is accurate and complete, check this box ☐ and return form to the address shown at the top of this notice. IF INACCURATE, CORRECT NAME AND OR NUMBER AND RETURN NOTICE.

If you have, or should have, accounts for business taxes not indicated above, please provide the information on those additional accounts. Use the back of this form if additional space is needed.

Other Identification Numbers	Other Account Name(s)	Corporation Tax	Sales & Use Tax	Withholding Tax

Return any notice(s) you receive to the above address within 10 days of receipt to avoid the possibility of any penalty.
If you receive more than one notice, enter complete information on one notice, attach all others, and return to the address shown above.

I certify that the above information is correct.

Business Phone and Area Code | Signature | Title | Date

DTF-132 (9/84)

3B 319,412

By now the public issue of Earthworm was well along, and it occurred to me that Goodale had not responded to my letter of October 10th.

RATERPILLAR TRACTOR CO.

495 Ashford Avenue
Ardsley, New York 10502
Telephone (914) 693-0400

December 4, 1984

Mr. Eugene Goodale,
Supervising Patent Attorney
Caterpillar Tractor Company
100 N.E. Adams Street
Peoria, IL 61629

Dear Eugene:

I have not heard from you for some time.

Cat got your tongue?

As I know you would expect of me as the President of Raterpillar to act in the best interests of its shareholder, I must point out that the costs incurred in operating this corporation continue. Since I last wrote you, I am certain that I have expended at least $1.20 in postage and further, I am now faced with the horrendous cost of paying my New York State Franchise Tax for 1985.

Should I be forced to do this, I feel that it would be necessary to withdraw my acceptance of your offer. With such a cost facing me, I will have no choice but to actively do business under the banner of the Rat. What's a Rat to do?

I know that things are hectic at your end, but if this becomes too complex a matter, you might pass it along to your Mergers and Acquisitions Committee for prompt and speedy action.

Cordially,

RATERPILLAR TRACTOR CO.

Michael Zinman,
President

MZ:llc

A Modest Disclaimer: Raterpillar Tractor Co. is in no way connected with Caterpillar Tractor Co. Caterpillar Tractor Co. sells "Cats" — Raterpillar Tractor Co. sells rats.

Cohen followed up the matter with a nudging letter to Robin.

LAW OFFICES OF

DECHERT PRICE & RHOADS

477 MADISON AVENUE
NEW YORK, NY 10022
TELEX 64 6324 • DARDEP

(212) 308-4400

3400 CENTRE SQUARE WEST
1500 MARKET STREET
PHILADELPHIA, PA 19102
(215) 972-3400

30 SQUARE DE MEEUS, STE 1
1040 BRUSSELS, BELGIUM
(02) 511 80 40

82 BEDFORD SQUARE
LONDON WC1B 3EX, ENGLAND
01. 631. 3383

1730 PENNSYLVANIA AVENUE, N.W.
WASHINGTON, DC 20006
(202) 783-0200

800 NORTH THIRD STREET
HARRISBURG, PA 17102
(717) 232-7947

ROBERT A. COHEN

January 7, 1985

Albert Robin, Esq.
Robin, Blecker & Daley
330 Madison Avenue (2nd Floor)
New York, New York 10017

Re: <u>Raterpillar Tractor Company</u>

Dear Al:

I hope we will be hearing from Caterpillar before it's time for the franchise tax returns to be filed.

Best regards.

Sincerely,

Robert

Robert A. Cohen

RAC:mc

bcc: Mr. Michael Zinman

Comes now one of those marvelous events in life that could never be planned. In early February, *Fortune* magazine publishes an issue featuring an article on barter and countertrade. Before reading it, you should know or remember that Caterpillar had just reported record losses of $450 million — its third year in a row — and cumulative losses totalling about one billion dollars. Business is really terrible for them at this time and there is a siege mentality all about.

They could not have come up with a more elegant product to barter. It was irresistible.

I wrote Mr. Schaefer a letter .

BETTER BARTER?

Basically a barter exchange, Univex lets corporate members use cash, too.

CATERPILLAR TRACTOR is working on a deal to sell tractors to a Latin American logger and sawmill operator. But since the logger will have trouble paying for them in dollars, Caterpillar is thinking about taking payment in coffins. That's one of barter's weakest points: the seller often has little use for the product the buyer offers.

A New York company called Universal Trading Exchange, Univex for short, is trying to get [illegible] by blending the worlds of barter and cash. Exchange members can trade products of unequal value and have the option—rare in the barter world—of settling their accounts in cash. No one gets stuck with unwanted goods.

One of 500 or so barter exchanges in the U.S., Univex has 40 members, including Ford, Du Pont, and TWA—but not Caterpillar. The board of directors includes Richard Shinn, former chairman of Metropolitan Life, and Walter Wriston, former Citicorp chairman. "I wanted instant credibility," says Univex Chairman Philip Stein, 54, a lawyer who founded the company in 1983 with $500,000 from Citicorp Venture Capital.

The exchange plans to make money on membership dues—$10,000 the first year and $5,000 a year thereafter for most members—and on transaction fees. For example, E. Leitz, the U.S. distributor of Leica cameras, recently bought carpet from Burlington Industries and gave Univex the standard 6% transaction fee as a cash deposit. Burlington receives trade credits it can use to buy goods from anyone in the exchange. If no one in the exchange makes anything Burlington wants, the textile company can take the remaining 94% in cash at the next biannual settlement time. Since the IRS counts trade credits as income, settlement in goods or services offers no tax advantage over cash.

Univex appeals to sellers because it matches them up with companies they might otherwise never do business with. Burlington never supplied carpet to Leitz before getting into the exchange. In addition, the exchange helps companies unload excess products or, in the case of airlines, hotels, and car rental services, fill excess capacity by cutting prices without advertising the fact. The price cuts come from the delayed settlement dates fixed by the exchange. Purchasers don't pay for several months. That means Citicorp could buy notebooks from Ampad, an office supply company, and earn interest before parting with the cash. The discount is about equal to what Citicorp can earn on the money.

Since Univex just started trading last March, it hasn't pulled in much in membership or transaction fees. It lost about $1 million on revenues of only $325,000 for the first nine months of 1984. Still, investors seem high on the operation. The exchange, which went public in late 1983, outdistanced all but one other publicly traded U.S. company last year; its stock, which trades over the counter, rose 484% in 1984 and recently sold at $10.50 a share. (The No. 1 highflier, Staar Surgical, also sold over the counter, lost money last year too.)

Trading possibilities through Univex seem limited only by a member's imagination and tolerance for rigmarole. In one of the more convoluted trades in the works, Playboy Enterprises Inc. is negotiating to buy a $750,000 telephone switching system from United Technologies. With its trade credits and cash, United Technologies will buy seats on TWA and rooms at Holiday Inns. Coffins could be a harder sell. **—Jaclyn Fierman**

RATERPILLAR TRACTOR CO.

495 Ashford Avenue
Ardsley, New York 10502
Telephone (914) 693-0400

February 4th 1985

Mr. George Schaefer
Chairman
Caterpillar Tractor Co.
Peoria, Illinois 61629

Dear Mr. Schaefer,

What with being new in the job, and one not without problems, I certainly don't wish to add to them, but feel I must bring to your attention a pending matter of long standing.

I have had some business dealings with your company regarding the sale of Raterpillar to Caterpillar, but it has been many weeks now since I last heard from your legal department who was to send me the necessary agreements of purchase and sale, and although I had come to believe that Cats word was its bond, its actions, or rather lack of action regretably would seem to belie this. It may be that your legal department is just buried in work, but I don't know.

I thought we had a deal.

My lawyers (you know what they are) tell me we have a deal - an enforceable one at that, and have suggested a course of action (you know what that would be) that I am personally loathe to embark on, but your failure to follow through in concluding our transaction limits my choice of action.

A temporary solution comes to mind. While in no way retreating from our deal, Raterpillar could do some business to at least cover its modest overhead. What with the disclosure in Fortune of Cats secret expansion plans, I propose you license Raterpillar to use the Cat logo on a line of coffins. I have in mind a new approach to the business - decorator colors. I don't know if yellow goes with everything, but we can try it.

Even if this endeavor is unsuccessful, I thought a Christmas type gift - a three inch coffin for the desk, emblazoned with the Cat "C", would be a nifty and fast selling item, and would constantly remind the Cat customer that Cat is always thinking of him. We could even personalze the lid if you thought it not in too poor taste.

How do you feel about this? What do you think the royalties should be for each coffin? Would you retain residual rights to the logo in the used coffin market?

A Modest Disclaimer: Raterpillar Tractor Co. is in no way connected with Caterpillar Tractor Co. Caterpillar Tractor Co. sells "Cats" — Raterpillar Tractor Co. sells rats.

RATERPILLAR TRACTOR CO.

495 Ashford Avenue
Ardsley, New York 10502
Telephone (914) 693-0400

Mr. George Schaefer
February 4th 1985
page -2-

How do you think your authorized dealers would feel about this?

As I explained to Mr. Thornton, Raterpillar would never sell tractors, but coffins - that's another matter. The way things are going in Peoria, we could probably set up shop and presell a lot of them - to current or recently current employees. We could even do a special line for management - producing these in Belgium and France - a nice touch, I would think.

Quick! Send me your standard royalty application form. Let's strike while the iron is hot.

I do hope we hear from you promptly with either the papers to close the deal on Raterpillar or the licensing agreement for the coffins.

Good luck in your new job.

Sincerely,

RATERPILLAR TRACTOR CO.

Michael Zinman
President

nm

P.S. Have you projected how many cents per share this coffin venture will contribute to Cat's earnings in 1985 and 1986? Do let me know - I'm a stockholder.

A Modest Disclaimer: Raterpillar Tractor Co. is in no way connected with Caterpillar Tractor Co. Caterpillar Tractor Co. sells "Cats" — Raterpillar Tractor Co. sells rats.

Judith Keller is Al Robins' secretary. I did not know it when I received the letter and was really quite puzzled. I phoned her and she told me that she in fact worked for Al and thought the entire matter hilarious.

I wrote her back just in case I was being set up. [Paranoia sets in after a while, and we were almost at the end of the saga].

21A Pilgrim's Harbor
Wallingford, Connecticut 06492
February 14, 1985

Mr. Michael Zinman, President
Raterpillar Tractor Co.
495 Ashford Avenue
Ardsley, New York 10502

Dear Mr. Zinman:

It has recently come to my attention that you are offering for sale three inch coffins in my favorite color - yellow. I am very interested in purchasing a quantity (depending on price) of these coffins.

Please rush me any information, catalogs or brochures you have on this product as I am currently holding in my refrigerator for burial twenty (possibly more) dead or dying earthworms.

Sincerely yours,

Judith Keller

JK:lb

RATERPILLAR TRACTOR CO.

495 Ashford Avenue
Ardsley, New York 10502
Telephone (914) 693-0400

February 19th 1985

Ms Judith Keller
21A Pilgrim's Harbor
Wallingford, Conn 06492

Dear Ms Keller,

In reply to your letter of February 14th, I must regretfully inform you that your information is faulted (as so is life).

We are <u>not</u> offering for sale three inch coffins. We are attempting to secure certain marketing rights for a particular line of coffins that may be offered for sale (according to Fortune Magazine) by the Caterpillar Tractor Co.

In line with this, we suggested to Caterpillar that they allow us to produce a desk top model as sort of a remembrance of things past.

We are anxiously awaiting their decision on this matter, and should it be favorable, we will revert to you with alacrity.

Keep the Earthworms crawling.

Formally yours,

RATERPILLAR TRACTOR CO.

Michael Zinman
President

nm

A Modest Disclaimer: Raterpillar Tractor Co. is in no way connected with Caterpillar Tractor Co. Caterpillar Tractor Co. sells "Cats" — Raterpillar Tractor Co. sells rats.

I chanced in on Bob Cohen one fine day in March, and he handed me the following memo that he had just received.

It speaks volumes.

FILE NO.

To: RAC

M. R Goodale Phone No.

Of: 309-675-5089

TELEPHONED	
CAME IN	
RETURNED CALL	
PLEASE RETURN CALL	
WILL CALL AGAIN	
WANTS TO SEE YOU	
DISCUSS WITH ME	
URGENT	

MESSAGE:

Re: ZINMAN-Raterpillar

Mr. Goodale Said that Bob Thornton will draft the papers + get them out to you later this week.

Any questions, please call

Date: Time: By: ☐ Read

\# 48 TELEPHONE MESSAGE ALL-STATE LEGAL SUPPLY CO., CRANFORD, N.J. 07016

We really were taking no chances. Caterpillar had agreed to raise the ante to $500 provided we did in fact pay the franchise tax we had told them was due. If for some reason the deal blew apart, we wanted continuity of the Corporation.

For the record, the relevant tax documents are included. I don't know how carefully the Government reads these returns, but I doubt if they ever had an official title of "Primary Rat" or "Principal Rat".

LAW OFFICES OF

DECHERT PRICE & RHOADS

477 MADISON AVENUE

NEW YORK, NY 10022

TELEX 64 5224 • BARDEP

(212) 308-4400

3400 CENTRE SQUARE WEST
1500 MARKET STREET
PHILADELPHIA, PA 19102
(215) 972-3400

36 SQUARE DE MEEUS, STE 1
1040 BRUSSELS, BELGIUM
(02) 511 80 40

52 BEDFORD SQUARE
LONDON WC1B 3EX, ENGLAND
01. 631. 3383

1730 PENNSYLVANIA AVENUE, N.W.
WASHINGTON, DC 20006
(202) 783-0200

800 NORTH THIRD STREET
HARRISBURG, PA 17102
(717) 238-7047

ROBERT A. COHEN

March 11, 1985

Mr. Michael Zinman
Earthworm Tractor Co. (Inc.)
495 Ashford Avenue
Ardsley, New York 10502

Re: Raterpillar

Dear Michael:

Here is the 1984 corporation franchise tax "reports and instructions." As I mentioned, the return is due March 15th. Mr. Atterbury has agreed that Caterpillar will bear the cost of the franchise tax ($250), and that he is sending out the papers on the sale today.

Best regards.

Sincerely,

Robert

Robert A. Cohen

RAC:mc
Encl.

CT-4

New York State Department of Taxation and Finance

Corporation Franchise Tax Report

Article 9-A, Tax Law

For calendar year **1984**
or taxable period
beginning ________ •
ending ________ •

S corporation must attach a copy of all pages of its federal Form 1120S *(Please print or type)*

Official use only

Employer identification number | File number

Attach Mailing Label Here →

If label is wrong, please correct it

Name CT-1556754 AA06 003077 DEC
CATERPILLAR TRACTOR CO., INC.
Num DECHERT ETAL
477 MADISON AVE
City NEW YORK NY 10022

Date received

Principal business activity | Telephone number | Business group code number from federal return •

State or country of incorporation: New York Date 1/84 | Date began business in New York State: 1/24/84

Audit use only

Federal return was filed on
☐ 1120 ☐ 1120-A ☐ 1120-S ☐ Consolidated basis ☐ Other ________

Have you been audited by the Internal Revenue Service in the past 5 years?
☐ No ☐ Yes If yes, list years

Total assets, line 30, Col. b: $100 •

IMPORTANT: To avoid rejection of form, complete all lines marked • Use zeros where they apply.

PAYMENT

Pay amount shown at line 24, Schedule A. Make check payable to: New York State Corporation Tax → $ 250 •

SCHEDULE A - COMPUTATION OF INCOME AND TAX

1. Federal taxable income before net operating loss deduction and special deductions		1	$ NONE •
2. Interest on federal, state, municipal and other obligations not included in line 1		2	•
3. New York State franchise tax deducted on federal return		3	•
4. Interest paid to stockholders: $______ • less 10% or $1000, whichever is larger *(see instructions)*		4	•
5. ACRS depreciation deducted in the computation of federal taxable income		5	•
6. Add lines 1, 2, 3, 4 and 5		6	NONE
7. Less - Depreciation allowed by Section 208.9(j) *(attach Form CT-399)*		7	•
8. Taxable income *(subtract line 7 from line 6)*		8	NONE •
9. Tax based on income	line 8 x 10% •	9	NONE •
10. Tax based on capital *(enter from Schedule C, line 32)*	x .00178 •	10	NONE •
11. Compensation of officers *(enter from Schedule D, line 37)*	Professional corporations enter zeros in lines 11 through 17.	11	•
12. Taxable income *(from line 8 above)*		12	
13. Add lines 11 and 12		13	NONE
14. Statutory exemption of $30,000 *(see instructions)*		14	
15. Subtract line 14 from line 13		15	NONE
16. Alternative base *(multiply line 15 by 30%)*		16	
17. Tax based on alternative base	line 16 x 10% •	17	NONE •
18. Minimum tax *(see instructions)*	$250.00	18	250
	TAX CREDITS CAN NOT BE CLAIMED ON THIS FORM - USE FORM CT-3		
19. Tax *(Enter largest of lines 9, 10, 17 or 18)*		19	250 •
20. Prepayments - see back		20	•
21. Balance *(Subtract line 20 from line 19)*		21	250
22. Interest: compute on line 21 amount *(see instructions)*		22	•
23. Additional charges: compute on line 21 amount		23	•
24. **BALANCE DUE** *(Add lines 21, 22 and 23)*	PAY →	24	250 •
25. **OVERPAYMENT** *(Subtract line 19 from line 20)*	CREDIT to next period →	25a	$ •
	REFUND →	25b	$ •

SCHEDULE B - ADDITIONAL REQUIRED INFORMATION

26. Compensation of officers deducted on federal return	26	NONE
27. Interest deducted on federal return	27	
28. Depreciable assets and land	28	
29. Total receipts *(total income plus cost of goods sold from federal return)*	29	NONE

SCHEDULE C - COMPUTATION OF CAPITAL

	a. Beginning of year	b. End of year	c. Average fair market value (see instructions)
30. Total assets *(from balance sheet of federal return)*	$ —	$ 100	$ 50
31. Current liabilities *(see instructions)*			—
32. Total capital *(subtract line 31 from line 30)* - Enter on line 10, Schedule A →			50
33. Issued capital stock			100
34. Paid-in capital, surplus and reserves			

35. Corporations organized outside New York State should complete the following for capital stock issued and outstanding:

Par Shares, $; No par Shares, $

36a. Does this corporation have an interest in real property located in New York State? ☐ Yes ☑ No

b. Has the controlling interest in the corporation's stock changed during the period covered by this report? ☐ Yes ☐ No
If both questions were answered "yes" attach a rider with complete details. See instructions.

SCHEDULE D - OFFICERS AND STOCKHOLDERS - List all officers, whether or not receiving any compensation, and every stockholder owning more than 5% of taxpayer's issued capital stock who received any compensation. It is mandatory to list the name, address and social security number of each officer.

Name and Address - Give actual residence (Attach rider if necessary)	Social Security Number	Official Title	Salary and All Other Compensation Received from Corporation
MICHAEL ZINMAN	107-V6-3108	PRESIDENT	NONE

37. Total compensation of officers *(including any amount on rider)* - Enter also on line 11, Schedule A → NONE

COMPOSITION OF PREPAYMENTS CLAIMED AT LINE 20, SCHEDULE A

	DATE	AMOUNT	DEPOSIT SERIAL NUMBER
a. Mandatory first installment		$	
b. CT-400 installments (1)			
(2)			
(3)			
c. Payment with extension application, Form CT-5			
d. Credit from prior years			
e. Total *(add lines a, b, c and d; enter on line 20)*		$	

CERTIFICATION BY AN ELECTED OFFICER OF THE CORPORATION

I hereby certify that this report, including any accompanying rider, is to the best of my knowledge and belief a true, correct and complete report.

3/15/85 — Date | [signature] — Signature of officer | PRIMARY RAT — Official title

Date | Signature of individual or name of firm preparing this report | Preparer's address

Mail this report to: Processing Unit
P.O. Box 1909
Albany, New York 12201

CT-3M/4M

New York State Department of Taxation and Finance

For calendar year 1984
or taxable period
beginning ____________
ending ____________

Metropolitan Transportation Business Tax Surcharge Report

Article 9-A, Section 209-B

Official use only

Date received

Attach Mailing Label Here →

Employer identification number	File number
Name	
Number	
City	

CT-1556754 AA06 002890 DEC
CATERPILLAR TRACTOR CO., INC.
DECHERT ETAL
477 MADISON AVE
NEW YORK NY 10022

Mail to:
Processing Unit
P.O. Box 1909
Albany, N.Y. 12201

ALL corporations required to file Forms CT-3, CT-3A or CT-4 **MUST** complete this form.

Does the above named corporation (or if filing on a combined basis, any corporation included in the combined group) do business, employ capital, own or lease property or maintain an office in the Metropolitan Commuter Transportation District, which includes the counties of New York, Bronx, Kings, Queens, Richmond, Dutchess, Nassau, Orange, Putnam, Rockland, Suffolk and Westchester?

☑ Yes • ☐ No •

If you answered yes, complete this form and pay the tax surcharge.

If you answered no, it is not necessary to complete Schedules N or O. However, this form **MUST** be signed by an officer and returned to: **Processing Unit, P. O. Box 1909, Albany, NY 12201**

		PAYMENT
Remit amount shown at line 9, Schedule N - Make check payable to N.Y. State Corporation Tax		$

SCHEDULE N — Computation of Tax Surcharge

1. Net New York State franchise tax from Form CT-3, CT-3A or CT-4	1	$ 250
2. M.C.T.D. Allocation percentage from Schedule O, line 15	2	100.0 %
3. Allocated tax — multiply line 1 by line 2	3	$ 250
4. **TAX SURCHARGE** — Multiply line 3 by 17%	4	$ 43
5. Prepayments (a) Payment with CT-5 .... $		
(b) Credit transferred from Form CT- ____ PERIOD ____ $		
Total Prepayments claimed	5	$
6. Balance - subtract line 5 from line 4	6	$ 43
7. Interest	7	$
8. Additional charges	8	$
9. **BALANCE DUE** - Add lines 6, 7 and 8	9	$ 43
10. Overpayment - Subtract line 4 from line 5		
a. Refund of overpayment	10a	$
b. Credit to N.Y. State franchise tax CT- ____ PERIOD ____	10b	$

CERTIFICATION OF AN ELECTED OFFICER OF THE CORPORATION

I hereby certify that this report, including any accompanying rider, is to the best of my knowledge and belief a true, correct and complete report.

3/15/85	[signature]	PRIMARY RAT
Date	Signature of officer	Official title
Date	Signature of preparer or name of firm	Preparer's address

SCHEDULE O — Computation of M.C.T.D. Allocation Percentage

	COLUMN A - M.C.T.D.	COLUMN B - N.Y. STATE
11. Average value of: *(Value property owned by same method as Sch.G, Form CT-3)*		
(a) Real estate owned		
(b) Real estate rented - (attach rider)		
(c) Inventories owned		
(d) Other tangible personal property owned		
(e) Total	$	$
(f) Percentage in M.C.T.D. (Divide Column A by Column B)		%
12. Receipts, in the regular course of business, from:		
(a) Sales of tangible personal property shipped to points within M.C.T.D.		
(b) All sales of tangible personal property		
(c) Services performed		
(d) Rentals of property		
(e) Royalties		
(f) Other business receipts		
(g) Total	$	$
(h) Percentage in M.C.T.D. (Divide Column A by Column B)		%
13. Wages, salaries, and other compensation of employees, except general executive officers	$	$
(a) Percentage in M.C.T.D. (Divide Column A by Column B)		%
14. Total of M.C.T.D. percentages shown at lines 11(f), 12(h) and 13(a)		%
15. M.C.T.D. ALLOCATION PERCENTAGE *(divide line 14 by 3 or by the number of percentages)*		%

INSTRUCTIONS

General Information

Section 209-B was amended by Chapter 999 of the Laws of 1984. This amendment extended the business tax surcharge on corporations doing business in the Metropolitan Commuter Transportation District for calendar years 1984 and 1985 and fiscal years ending in 1985 and 1986. The M.C.T.D. includes the counties of New York, Bronx, Queens, Kings, Richmond, Dutchess, Nassau, Orange, Putnam, Rockland, Suffolk and Westchester.

Who Must File

All corporations taxable under Article 9-A of the Tax Law. This includes taxpayers that file Forms CT-3, CT-3A and CT-4.

Corporations filing on a combined basis are required to file only one report for the combined group. Combined figures, as shown on your CT-3-A, should be used in completing this form.

Who Must Pay the Tax Surcharge

Every corporation taxable under Article 9-A that is doing business, employing capital, owning or leasing property or maintaining an office in the Metropolitan Commuter Transportation District (See General Information) must pay a business tax surcharge.

A corporation is not subject to this business tax surcharge if its activities are limited to solicitation of orders by the corporation's representatives or independent contractors for sales of tangible personal property, which orders are sent outside the district for approval or rejection, and, which if approved, are filled by shipment or delivery from a point outside the district.

Tax Rate

18% for calendar year 1982 or fiscal years ending in 1983

17% for calendar year 1983, 1984 and 1985 or fiscal years ending in 1984, 1985 and 1986

The surcharge rate is applied to the portion of tax, after deduction of any credits, imposed under Section 209 attributable to the taxpayer's business activity carried on within the M.C.T.D.

Time for Filing

Two and one-half months after the close of the reporting period.

Payment of Tax Surcharge

The entire tax surcharge must be paid on or before the original due date of the report.

Extension of Time for Filing Tax Surcharge Report

A request for an extension of time to file a tax surcharge report must be filed on Form CT-5 on or before the due date of the report, for the taxable period for which the extension is requested.

Requirements for a valid extension for all MTA tax surcharge reports are as follows:

a. A separate application for extension, along with any estimated tax surcharge payments that may be due, must be filed on or before the due date of the report for the taxable period for which this extension is requested,

and

b. The total tax surcharge payment for the period for which this extension is requested must equal 90% of the tax surcharge as finally determined.

Specific Instructions

All corporations that file Form CT-3, CT-3A or CT-4, must answer yes or no to the question preceding Schedule N. **It is essential that corporations not doing business in the Metropolitan Commuter Transportation District disclaim liability for the tax surcharge by answering "no."**

Schedule N — Computation of Tax Surcharge

Line 1. Enter your net franchise tax from your corporation franchise tax report.

Form CT-3, Schedule A, Line 6c

Form CT-3A, Schedule H, Line 6e

Form CT-4, Schedule A, Line 19

Line 2. Enter your M C T D allocation percentage from Schedule O, line 15.

Line 5a. Enter any payment made with Form CT-5, Application for Three-Month Extension.

Line 5b. You can apply an overpayment of franchise tax, made on your New York State franchise tax report to pay this tax surcharge. Enter at this line the form number, period in which the overpayment occurred (use month and year) and amount to be applied. At the overpayment line of your New York State franchise tax report indicate the amount of money to be applied to the M.C.T.D. tax surcharge.

Lines 7 and 8 Interest and additional charges — — See Forms Booklet CT-2, for instructions.

Schedule O — Computation of M.C T.D. allocation percentage.

The M.C.T.D allocation is determined by averaging the percentages of the corporation's property, receipts and wages that apply to the Metropolitan Commuter Transportation District.

- A corporation that does all of its New York State business within the 12 counties of the M.C.T.D. can enter 100% at line 15 and at line 2 of Schedule N.
- A corporation that does part of its business outside of the M.C.T.D., but not outside of New York State must compute its M.C.T.D. allocation by completing this schedule For detailed instructions, refer to the instructions for Schedule G, Form CT-3, found in your Corporation Franchise Tax Forms Booklet CT-2 When reading Schedule G instructions, substitute "New York State" for "Everywhere" and M.C.T.D. for New York State.
- A corporation that claimed a business allocation percentage by completing Schedule G, Form CT-3 must enter in Schedule O, Column B the NYS figures from Column A of Schedule G. The same instructions for Form CT-3, Schedule G apply, except you must substitute "New York State" for "Everywhere" and "M.C.T.D." for "New York State."

1234

Form **1120-A**

Department of the Treasury
Internal Revenue Service

U.S. Short-Form Corporation Income Tax Return

To determine whether you qualify to file Form 1120-A, see Instructions.

For calendar 1984 or tax year beginning 1/4, 1984, ending 12/31, 19 84

▶ **For Paperwork Reduction Act Notice, see page 1 of the Instructions.**

OMB No 1545-0890

1984

See Instructions for list of principal business: A Activity; B Product or service; C Code

Use IRS label. Otherwise please type or machine print.

Name: RATERPILLAR TRACTOR CO, INC

Number and street: C/O DECHERT PRICE ET AL

City or town, State and ZIP code: 477 MADISON AVE NEW YORK, N.Y. 100VV

D Employer identification number (EIN)

E Date incorporated: 1/4/84

F Total assets (see Specific Instructions): $ 100

G Check method of accounting (1) ☐ Cash (2) ☐ Accrual (3) ☐ Other (specify) ▶

H Check box if there has been a change in address from the previous year ▶ ☐

Gross Income

1	(a) Gross receipts or sales ____ (b) Less returns and allowances ____ Balance ▶	1(c)	
2	Cost of goods sold and/or operations (see instructions)	2	
3	Gross profit (line 1(c) less line 2)	3	
4	Domestic corporation dividends subject to the 85% deduction	4	
5	Interest	5	
6	Gross rents	6	
7	Gross royalties	7	
8	Capital gain net income (attach separate Schedule D (Form 1120))	8	
9	Net gain or (loss) from Form 4797, line 14(a), Part II (attach Form 4797)	9	
10	Other income (see instructions)	10	
11	TOTAL income—Add lines 3 through 10	11	NONE

Deductions

12	Compensation of officers (see instructions)	12	
13	(a) Salaries and wages ____ (b) Less jobs credit ____ Balance ▶	13(c)	
14	Repairs	14	
15	Bad debts (If reserve method is used, answer Question K on page 2)	15	
16	Rents	16	
17	Taxes	17	
18	Interest	18	
19	Contributions (see instructions for 10% limitation)	19	
20	Depreciation (attach Form 4562) — 20		
21	Less depreciation claimed elsewhere on return — 21(a)	21(b)	
22	Other deductions (attach schedule)	22	
23	TOTAL deductions—Add lines 12 through 22	23	NONE
24	Taxable income before net operating loss deduction and special deductions (line 11 less line 23)	24	NONE
25	Less: (a) Net operating loss deduction (see instructions) — 25(a); (b) Special deductions (see instructions) — 25(b)	25(c)	
26	Taxable income (line 24 less line 25(c))	26	NONE

Tax

27	TOTAL TAX (from Part I, line 6 on page 2)	27	NONE
28	Payments: (a) 1983 overpayment allowed as a credit; (b) 1984 estimated tax payments; (c) Less 1984 refund applied for on Form 4466 ( ); (d) Tax deposited with Form 7004; (e) Credit from regulated investment companies (attach Form 2439); (f) Credit for Federal tax on gasoline and special fuels (attach Form 4136)	28	
29	Enter any PENALTY for underpayment of estimated tax—Check ▶ ☐ if Form 2220 is attached	29	
30	TAX DUE—If the total of lines 27 and 29 is larger than line 28, enter AMOUNT OWED	30	NONE
31	OVERPAYMENT—If line 28 is larger than the total of lines 27 and 29, enter AMOUNT OVERPAID	31	
32	Enter amount of line 31 you want Credited to 1985 estimated tax ▶ ____ Refunded ▶	32	

Please Sign Here

Under penalties of perjury, I declare that I have examined this return, including accompanying schedules and statements, and to the best of my knowledge and belief, it is true, correct, and complete. Declaration of preparer (other than taxpayer) is based on all information of which preparer has any knowledge.

▶ [signature] Signature of officer — Date 3/18/85 — ▶ PRINCIPAL RAT Title

Paid Preparer's Use Only

Preparer's signature ▶ — Date — Check if self-employed ▶ ☐ — Preparer's social security number

Firm's name (or yours, if self-employed) and address ▶ — E.I. No ▶ — ZIP code ▶

Form 1120-A (1984) **Part I** **Tax Computation (See Instructions)** Page 2 Enter EIN ▶

1 Income tax (see instructions to figure the tax; enter lesser of this tax or alternative tax from Schedule D) Check if from Schedule D ▶ ☐	1	NONE
2 General business credit. Check if from ☐ Form 3800 ☐ Form 3468 ☐ Form 5884 ☐ Form 6478 ☐ Form 8007	2	
3 Line 1 less line 2	3	
4 Tax from recomputing prior year investment credit (attach Form 4255)	4	
5 Minimum tax on tax preference items (see instructions—attach Form 4626)	5	
6 Total tax—Add lines 3 through 5. Enter here and on line 27, page 1	6	NONE

Additional Information (See instruction F)

I Was a deduction taken for expenses connected with:

(1) An entertainment facility (boat, resort, ranch, etc.)? Yes ☐ No ☑

(2) Employees' families at conventions or meetings? Yes ☐ No ☑

J Did any individual, partnership, estate or trust at the end of the tax year own, directly or indirectly, 50% or more of the corporation's voting stock? (For rules of attribution, see section 267(c).) If "Yes," complete (1) and (2) Yes ☐ No ☑

(1) Attach a schedule showing name, address, and identifying number

(2) Enter highest amount owed, include loans and accounts receivable/payable

(a) Enter highest amount owed by the corporation to such owner during the year ▶

(b) Enter highest amount owed to the corporation by such owner during the year ▶

K If the reserve method is used for bad debts, complete (1) and (2) for the current year

(1) Amount added to the reserve account

(a) Current year's provision ▶

(b) Recoveries ▶

(2) Amount charged against the reserve account ▶

L If an amount for cost of goods sold and/or operations is entered on line 2, page 1, complete (1) and (2)

(1) Purchases ▶

(2) Other costs (attach schedule) ▶

M At any time during the tax year, did you have an interest in or a signature or other authority over a bank account, securities account, or other financial account in a foreign country? (See instruction F for filing requirements for form TD F 90-22.1.) Yes ☐ No ☑

If "Yes," write in the name of the foreign country ▶

N During this tax year was any part of your accounting/tax records maintained on a computerized system? Yes ☐ No ☑

O Enter amount of cash distributions and the book value of property (other than cash) distributions made in this tax year ▶

Part II **Balance Sheets**

		(A) Beginning of tax year	(B) End of tax year
Assets	1 Cash		100
	2 [illegible]		
	(a) [illegible]		( )
	3 [illegible]		
	4 [illegible]		
	5 Other [illegible]		
	6 Loans [illegible]		
	7 Mortgage [illegible]		
	8 Depreciable, depletable and intangible assets		
	(a) Less accumulated depreciation, depletion and amortization	( )	( )
	9 Land (net of any amortization)		
	10 Other assets (attach schedule)		
	11 Total assets		100
Liabilities and Stockholders' Equity	12 Accounts payable		
	13 Other current liabilities (attach schedule)		
	14 Loans from stockholders		
	15 Mortgages, notes, bonds payable		
	16 Other liabilities (attach schedule)		
	17 Capital stock (Preferred and Common stock)		100
	18 Paid-in or capital surplus		
	19 Retained earnings		
	20 Less cost of treasury stock	( )	( )
	21 Total liabilities and stockholders' equity		100

COMPANY ORGANIZED 1/24/84

Part III **Reconciliation of Income Per Books With Income Per Return (See Instructions)**

1 Enter net income per books	NONE	5 Income recorded on books this year not included in this return (itemize)	
2 Federal income tax		6 Deductions in this tax return not charged against book income this year (itemize)	
3 Income subject to tax not recorded on books this year (itemize)		7 Income (line 24, page 1). Enter the sum of lines 1, 2, 3 and 4 less the sum of lines 5 and 6	NONE
4 Expenses recorded on books this year not deducted in this return (itemize)			

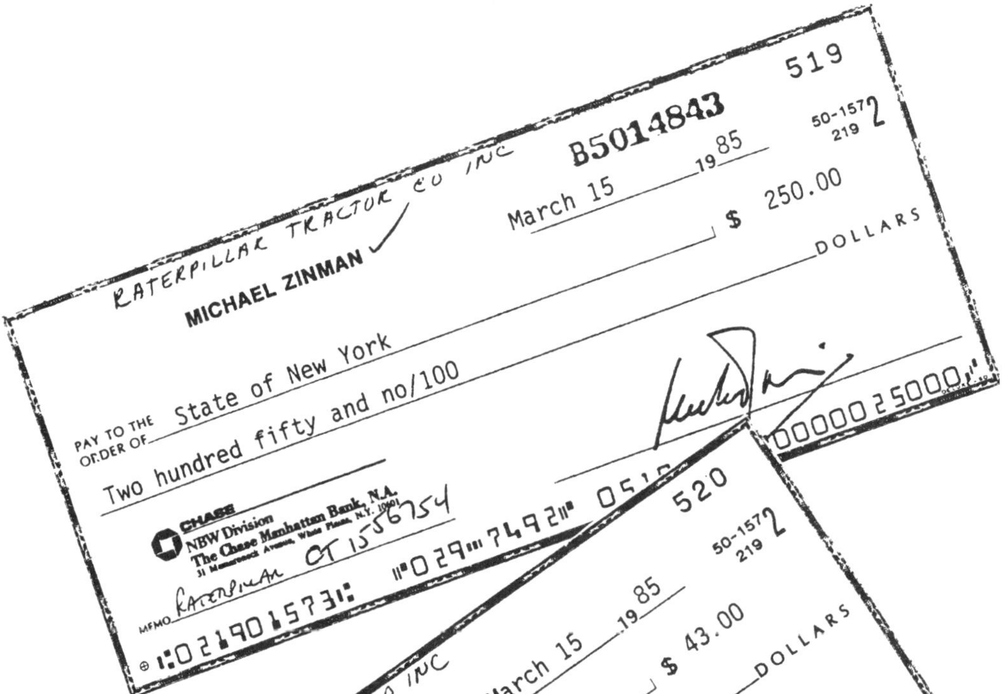

519

B5014843

50-157
219 2

RATERPILLAR TRACTOR CO INC

MICHAEL ZINMAN

March 15 19 85

$ 250.00

PAY TO THE ORDER OF State of New York

Two hundred fifty and no/100 DOLLARS

CHASE
NBW Division
The Chase Manhattan Bank, N.A.
31 Mamaroneck Avenue, White Plains, N.Y. 10601

CT 1556754

MEMO RATERPILLAR

⑆021901573⑆ ⑈029⑈7492⑈ 0519 ⑈0000025000⑈

520

50-157
219 2

RATERPILLAR TRACTOR CO INC

MICHAEL ZINMAN

March 15 19 85

$ 43.00

PAY TO THE ORDER OF State of New York

Forty three and no/100 DOLLARS

N5016484

CHASE
NBW Division
The Chase Manhattan Bank, N.A.
31 Mamaroneck Avenue, White Plains, N.Y. 10601

CT 1556754

MEMO RATERPILLAR

⑆021901573⑆ ⑈029⑈7492⑈ 0520 ⑈0000004300⑈

On March 11th, Caterpillar set the final wheels in motion and sent a draft agreement off to Bob Cohen. It totalled twenty-five pages.

In a peculiar way, this was Caterpillar's corporate sense of humor rising from the gorge. R. Rennie Atterbury, one of their senior counsel, dusted off an agreement they had used for a prior acquisition and sent it to Cohen. In it we were required to submit a number of supporting documents at the closing: a certificate stating that we were a corporation in good standing in the State of New York; audited financial statements; personal property statements, and other items. These can be examined in the complete closing document appended. Atterbury told me at the closing that they (Caterpillar) knew that by using this document and requiring me to supply all the exhibits, my legal costs would be pretty hefty—at least for what I was receiving for the company. I guess they were amused—for a while at least.

Cohen had a response drafted and sent it back to Atterbury on the 22nd of March. Caterpillar replied on April 1st. I signed some documents on the 6th, and all was ready for the closing on the 11th.

One Exhibit, "E," was held out. This was the "Personal Property" statement which was to include the inventory of Raterpillar: two rats.

I have not included the actual drafts that went back and forth to Peoria during this period as the final closing package is appended to this narrative. This document includes all of this material; to incorporate the drafts here would only add needless bulk.

CATERPILLAR TRACTOR CO.

Peoria, Illinois 61629-7310

March 11, 1985

Mr. Robert A. Cohen, Esq.
Dechert, Price & Rhoads
477 Madison Avenue
New York, N.Y. 10022

Dear Mr. Cohen:

Enclosed is a draft Share Purchase Agreement in connection with Caterpillar Tractor Co.'s proposed purchase of all of the outstanding shares of Raterpillar Tractor Co. from Mr. Zinman.

We assume you will have few difficulties with the enclosed draft, and that many, if not most, of the exhibits can be completed by the word "none".

Needless to say, we do not anticipate expending time or effort on "due diligence" type investigations or reviews and have prepared the agreement accordingly.

If you will provide me with any questions or comments you may have, and with the appropriate information or documents for inclusion in the various exhibits, we can finalize the agreement for signature by the parties and agree the date of the closing to be held in your offices.

Sincerely yours,

R. Rennie Atterbury

Associate Counsel

RRAtterbury III
Telephone: (309) 675-4477
lr

Atch.

rra5295130

How truly fitting

LAW OFFICES OF

DECHERT PRICE & RHOADS

477 MADISON AVENUE
NEW YORK, NY 10022
TELEX 64 6324 • BARDEP
(212) 308-4400

3400 CENTRE SQUARE WEST
1500 MARKET STREET
PHILADELPHIA, PA 19102
(215) 972-3400

30 SQUARE DE MEEUS, STE 1
1040 BRUSSELS, BELGIUM
(02) 511 60 40

32 BEDFORD SQUARE
LONDON WC1B 3EX, ENGLAND
01. 631. 3383

1730 PENNSYLVANIA AVENUE, N.W.
WASHINGTON, DC 20006
(202) 783-0200

800 NORTH THIRD STREET
HARRISBURG, PA 17102
(717) 233-7947

ROBERT A. COHEN

Federal Express

March 22, 1985

R. Rennie Atterbury, III, Esquire
Associate Counsel
Caterpillar Tractor Company
Peoria, Illinois 61629-7310

Re: Caterpillar Tractor Co. acquisition of Raterpillar Tractor Co.

Dear Mr. Atterbury:

Thanks for the proposed Share Acquisition Agreement which you prepared in connection with Caterpillar's proposed purchase of all the outstanding shares of Raterpillar from Michael Zinman. I've got a few thoughts about the Agreement and have enclosed a "marked-up" copy of the Agreement which shows my suggested modifications. Generally, the proposed changes serve to conform the Agreement more closely to the circumstances of the transaction. In particular, I propose the following modifications for the reasons indicated.

In Article 1, I think we should agree on a date certain for the Closing. I would suggest April 11, 12, 18 or 19, 1985. Let me know which date is convenient for you.

In Article 2, the changes reflect that fact that only one (1) share has been issued (conforming changes are made throughout), and provide for the issuance of a receipt for that share.

In Articles 3 and 7.4, the purchase price has been increased to Five Hundred Dollars ($500.00) to reflect the cost of the 1984 Corporate Franchise Tax in the amount of $293.00 recently paid by Michael Zinman on behalf of the Company and which you and I agreed would be borne by Caterpillar as part of the acquisition cost.

R. Rennie Atterbury, III, Esquire
Page two
March 22, 1985

In Articles 4.7, 4.17, 4.22, 4.24, 4.25, 4.26 and 4.27 I have modified the language so as to eliminate the need for otherwise superfluous exhibits to the Agreement. For the same purpose, I would recommend the following language to replace Article 4.23:

> 4.23 Industrial Property Rights. As of the date hereof, the Company does not possess nor does it have pending any application for United States or foreign patents, registered trademarks or trade names, unregistered trademarks or trade names, or copyrights or copyright registrations. Further the Company has never granted or been granted any license, or other agreement, relating in whole or in part to any industrial property rights, or to inventions, disclosures, improvements, processes, formulae, trade secrets, ideas or other know-how, or labels, whether or not patented or registered and whether owned by the Company or otherwise.

Also added to certain of the provisions in Article 4 is the term "material" with that term being defined as an amount in excess of Fifty Dollars ($50.00), paralleling your provision in Article 4.13.

As to our opinion as counsel to Seller, we are only able to opine to the matters set out in Articles 4.1, 4.2 and 4.4 to 4.9. You will, however, receive Michael Zinman's certificate as to all the provisions of Article 4.

In Article 10, I would eliminate the phrase "nor any person having a special relationship to Seller," since Michael Zinman is not in a position to obligate such persons to abide by such an agreement and his signature can not bind them.

I hope that you will not have any difficulties with these and the other modifications I've made which I think are self explanatory. If you should have any questions or comments, please call me to discuss them so that we can get the Agreement executed as soon as possible.

R. Rennie Atterbury, III, Esquire
Page three
March 22, 1985

I have also enclosed copies of my proposed Exhibits A through C for incorporation into the Agreement. Exhibits D and E are still in the works. I will send them to you as soon as they are available. Some proposed Closing documents, specifically the Closing Agenda and documents identified in items A-1, A-5, A-6, B-1 and C therein, are also enclosed for your review. I look forward to meeting you at the Closing.

Sincerely,

Robert A. Cohen

RAC:mc
Encls.

cc: Mr. Michael Zinman
(w/encls.)

LAW OFFICES OF

DECHERT PRICE & RHOADS

477 MADISON AVENUE

NEW YORK, NY 10022

TELEX 64 5884 • BARDEP

(212) 308-4400

3400 CENTRE SQUARE WEST
1500 MARKET STREET
PHILADELPHIA, PA 19102
(215) 972-3400

36 SQUARE DE MEEUS, STE 1
1040 BRUSSELS, BELGIUM
(02) 511 60 40

52 BEDFORD SQUARE
LONDON WC1B 3EX, ENGLAND
01. 631. 3363

1730 PENNSYLVANIA AVENUE, N.W.
WASHINGTON, DC 20006
(202) 783-0200

800 NORTH THIRD STREET
HARRISBURG, PA 17102
(717) 233-7947

ROBERT A. COHEN

April 3, 1985

Mr. Michael Zinman
Earthworm Tractor Co. (Inc.)
495 Ashford Avenue
Ardsley, New York 10502

Re: Raterpillar Tractor Co.

Dear Michael:

For the book, enclosed is a copy of Atterbury's recent letter and its enclosure.

We still need Exhibit D which contains the financial statements referred to in paragraph 4.10 and Exhibit E which is a list of personal property owned by the Company referred to in paragraph 4.15.

Best regards.

Sincerely,

Robert

Robert A. Cohen

RAC:mc
Encl.

CATERPILLAR TRACTOR CO.

Peoria, Illinois 61629-7310

April 1, 1985

Mr. Robert A. Cohen
Dechert Price & Rhoads
477 Madison Avenue
New York, N.Y. 10022

Re: Raterpillar Tractor Co.

Dear Mr. Cohen:

Pursuant to our telephone conversation last Friday, enclosed is a copy of the revised Share Acquisition Agreement. I presume you will find it acceptable.

Exhibits A, B and C enclosed with your March 22 letter are acceptable to us. I would appreciate receiving copies of Exhibits D and E at your earliest convenience.

As soon as all exhibits have been agreed, we can arrange for the Share Acquisition Agreement to be signed by the parties.

We understand the Closing will take place in your offices at approximately 1:30 p.m. on April 11, 1985. Assuming Raterpillar Tractor Co. has no books of account or other financial records, the Closing Agenda proposed by you appears acceptable -- except that Caterpillar's name should be "Caterpillar Tractor Co.", not "Company."

In view of the magnitude of this transaction, I shall be the sole representative of Caterpillar attending the Closing. I look forward to meeting you and Mr. Zinman at that time.

Sincerely,

R. Rennie Atterbury

Associate Counsel

RRAtterbury III
Telephone: (309) 675-4477
lr

Enc.

rra5274230

LAW OFFICES OF

DECHERT PRICE & RHOADS

477 MADISON AVENUE
NEW YORK, NY 10022
TELEX 64 6324 • BARDEP

(212) 308-4400

3400 CENTRE SQUARE WEST
1500 MARKET STREET
PHILADELPHIA, PA 19102
(215) 972-3400

36 SQUARE DE MEEUS, STE 1
1040 BRUSSELS, BELGIUM
(02) 511 80 40

52 BEDFORD SQUARE
LONDON WC1B 3EX, ENGLAND
01. 631. 3383

1730 PENNSYLVANIA AVENUE, N.W
WASHINGTON, DC 20006
(202) 783-0300

800 NORTH THIRD STREET
HARRISBURG, PA 17102
(717) 233-7947

April 5, 1985

FEDERAL EXPRESS

Mr. Michael Zinman
Earthworm Tractor Co. (Inc.)
495 Ashford Avenue
Ardsley, New York 10502

Re: Raterpillar Tractor Co.

Dear Mr. Zinman:

Enclosed are two copies of the Share Acquisition Agreement together with Exhibits A, B, C and E. Please put the "finishing touches" on both copies as indicated below:

1) Sign and date the Certificates of the Secretary which are found on the first pages of Exhibits A and B. The Corporate Seal is already affixed.

2) Insert the Financial Statements and the Accountant's Certificate as Exhibit D.

3) Fill in the cost of the "personal property" in the column labelled "Value" on Exhibit E.

4) Sign and date the Acquisition Agreement and initial each page thereof.

5) Send the executed copies of the Agreement, the Exhibits and the enclosed cover letter by Federal Express to:

Mr. R. Rennie Atterbury, Esquire
Associate Counsel
Caterpillar Tractor Co.
Peoria, Illinois 61629-7310

Mr. Michael Zinman
Page Two
April 5, 1985

Hopefully, this should do it until Thursday. I look forward to seeing you and the "personal property" then.

Sincerely,

Karl Okamoto
Law Clerk

KO:bf
Enclosures

About the rats.

I thought it would be difficult to purchase these items but knew of a laboratory breeding company in Massachusetts. I had some difficulty ordering them, and subsequently cancelled the order when I found that rats are a staple of large snake's diets, and, as such, are available in many pet stores in the area.

On the day of the closing, I purchased two of them. I named them Lee and Chappie.

RATERPILLAR TRACTOR CO.

495 Ashford Avenue
Ardsley, New York 10502
Telephone (914) 693-0400

March 15th 1985

Ms Sheila Loranger
Charles River Breeding Laboratories
251 Ballardvale Street
Wilmington, Mass 01887

Dear Ms. Loranger,

Please find enclosed our purchase order No. 1 covering two white rats we are buying from your firm.

We had hoped to be a really big user of the rats, as we felt we could do a bang up job selling them, but it seems that Caterpillar Tractor Co. wants to take this business to themselves, and so have made us an offer we can't refuse.

As we are planning to sell our company to Caterpillar (honest!) we felt they should have a start in whatever they plan to do with it, so we want to provide them with some breeding stock.

We think they may not want to do much breeding of rats, and so you might contact them in the near future and determine what plans they have in the rat market.

You may land a totally unexpected new customer who can blaze trails in new fields of endeavors.

Good luck to you both.

Very truly yours,
RATERPILLAR TRACTOR CO. INC

Michael Zinman
President

nm
enc

A Modest Disclaimer: Raterpillar Tractor Co. is in no way connected with Caterpillar Tractor Co. Caterpillar Tractor Co. sells "Cats" — Raterpillar Tractor Co. sells rats.

RATERPILLAR TRACTOR CO.

495 Ashford Avenue
Ardsley, New York 10502
Telephone (914) 693-0400

PURCHASE ORDER

Charles River Breeding Laboratories
251 Ballardvale Street
Wilmington, Mass 01887
Attention: Sheila Loranger

DATE Mar 15th 1985
OUR P.O. NO. One
SHIP VIA Best Way
F.O.B. Peoria, Ardsley or New York
To Be Advised
SALESMAN Zinman

NOTE: Purchase order number must appear on all invoices

SHIP TO:

To Be Advised

DESCRIPTION	AMOUNT
Two (2) Genuine American White Rats (Genus Rattus), one Male one Female.	
Price, each $20.00	$40.00

Any issue from this pair becomes the property of Raterpillar Tractor Co (or its assigns).

Charles River Breeding Laboratories guarantees Rats will be alive when shipped.

Shipping Charges will be prepaid by Charles River and are included in the price.

Terms of Payment: Net Cash, check enclosed
Please acknowledge and forward paid invoice upon receipt of this purchase order.

RATERPILLAR TRACTOR CO, INC

A Modest Disclaimer: Raterpillar Tractor Co. is in no way connected with Caterpillar Tractor Co. Caterpillar Tractor Co. sells "Cats" — Raterpillar Tractor Co. sells rats.

The closing was professionally a model—everything in its place and everyone appearing to know exactly what needed doing. It was a warm spring day, and the sun was shining. All was in order in the large conference room at Dechert Price and Rhoads. When I arrived (with Schedule "E"), we immediately began signing the many papers.

Atterbury would not take the rats. He said that he had studied to be a veterinarian and knew they would soon gnaw their way out of the box. On closing, all assembled were my treat to dinner, and while Atterbury and I were waiting in the lobby of the law firm, I with the rat box under my arm, one of the secretaries asked to see the rats, and fell in love with the little creatures—honest. (She told me later that she was partial to Lee). I offered, she accepted, and Atterbury and I beat a hasty exit to a waiting elevator.

A few loose ends were completed the next day by Karl Okamoto of Dechert (which included the only factual statement as to the disposition of the rats), and all that was left to do to complete the story was to place the tombstone ad in *The Wall Street Journal.*

Alas, this was not to be.

3400 CENTRE SQUARE WEST
1500 MARKET STREET
PHILADELPHIA, PA 19102
(215) 972-3400

36 SQUARE DE MEEUS, STE 1
1040 BRUSSELS, BELGIUM
(02) 511 80 40

52 BEDFORD SQUARE
LONDON WC1B 3EX, ENGLAND
01 631 3383

DECHERT PRICE & RHOADS
477 MADISON AVENUE
NEW YORK, NY 10022
TELEX 64 5324 · BARDEP

(212) 308 4400

[illegible] PENNSYLVANIA AVENUE, N.W.
WASHINGTON, DC 20006
(202) 783-0200

300 NORTH THIRD STREET
HARRISBURG, PA 17102
(717) 233-7947

April 12, 1985

R. Rennie Atterbury, Esquire
Associate Counsel
Caterpillar Tractor Co.
Peoria, Illinois 61629-7310

Re: Raterpillar Tractor Co., Inc.

Dear Mr. Atterbury:

It was very nice to have met you at our recent closing. Enclosed is the subsistence certificate for Raterpillar Tractor Co., Inc. which I promised to forward to you once it arrived from the Secretary of State for the State of New York.

The "personal property" of Raterpillar you left behind were released into their "natural habitat". Beth couldn't convince the home office to permit entry. Perhaps they're on their way to Peoria.

I hope that all went well in Japan and look forward to meeting you again.

Sincerely,

[signature]

Karl Okamoto
Law Clerk

On April 12th, I called *The Wall Street Journal* and asked how I should proceed to place an advertisement in the paper celebrating the sale of Raterpillar.

This caused unexpected problems. While initially accepting the copy, I received a call later in the afternoon and was told that a management decision had been made not to run the ad. In fact, they said they didn't believe me.

I sent a messenger from Dechert to *The Wall Street Journal* with all the closing documents, and while they were incredulous, they still wouldn't run it. They did, however, give me a few suggestions. They said they were sure *The New York Times* would run the ad (they were right—it did), and, more interestingly, suggested I put out a press release on the matter. They felt it quite newsworthy.

Well I did put out a press release, and the rest, as they say, is history. The story was picked up by the Gannett chain, the local Peoria paper, and the Associated Press. It also made the rounds of a number of cable television news services.

In due course, it came to the attention of Caterpillar. Some of the articles that appeared, with some of Caterpillar's P.R. comments, are included.

The New York Times, **Wednesday, April 17, 1985**

CATERPILLAR TRACTOR CO.

Has Acquired From

MICHAEL ZINMAN

All of the Outstanding Shares of

RATERPILLAR TRACTOR CO.

April 11, 1985

MICHAEL ZINMAN, Ardsley, New York

Contact: Robert Cohen, Esq
Dechert Price and Rhoads
477 Madison Avenue
New York, New York'
212-308-4400

FOR IMMEDIATE RELEASE

CATERPILLAR TRACTOR CO MAKES NEW ACQUISITION:

Ardsley, New York. April 11, 1985

Michael Zinman announced today the sale of Raterpillar Tractor Co to Caterpillar Tractor Co. The closing took place at the offices of his attorneys in New York City.

Incorporated in New York in 1984, Raterpillar sought to achieve for itself in the Rat Industry that which Caterpillar accomplished in the Construction Manufacturing Industry.

Although no sales figures were supplied, it was learned that the inventory of Raterpillar, consisting mainly of two rats, was turned over to Caterpillar at the closing. It was further learned that the rats then came into the possession of a secretary of the law firm, who was not able to be reached for comment as to her intentions with the furry animals.

more....

Zinman, President of Raterpillar at the time of the sale, commented that while rats were the main product that Raterpillar sought to market when the company was founded, consideration was given to market a line of coffins that Caterpillar was said to be receiving in a complex counterpart trade transaction they were involved in. This proposal, made by Raterpillar to Caterpillar, was unacted upon at the time of the sale of the company.

Zinman stated that Caterpillar had not advised him who they appointed President of this new subsidiary, but noted that as both E.C. Chapman and Lee Morgan had recently retired from high executive positions at Caterpillar, either might be a candidate under consideration for this position.

The purchase price was undisclosed but was reported to be in the mid to high three figures.

ooOoo

Walter Robinson, Esq. is a lawyer with a sense of humor.

This unusual man is a partner at Pillsbury Madison and Sutro in San Francisco. He was one of the main litigators in the case of *Caterpillar Tractor Co.* vs. *Earthworm Tractor Co.* (77 Civ 794 LPG).

WALTER J. ROBINSON
225 BUSH STREET
P.O. BOX 7880
SAN FRANCISCO, CA 94120

May 6, 1985

Michael Zinman
495 Ashford Avenue
Ardsley, NY 10502

Dear Michael:

On my return from a recent trip I found your gift. From your recent press, I assume that you are now involved totally with the animal world, and that your luck with ducks will be equal to your experience with rats and cats. In any case, I appreciate your obviously fond recollection of our prior involvement.

My regards to Bob Cohen and Mel Schwarz (whose toy collections must by now indeed be extraordinary).

Very truly yours,

Walt

Walter J. Robinson

The closing document file. . .

LAW OFFICES OF

DECHERT PRICE & RHOADS

477 MADISON AVENUE
NEW YORK, NY 10022
TELEX 84 6824 • BARDEP

(212) 308-4400

3400 CENTRE SQUARE WEST
1500 MARKET STREET
PHILADELPHIA, PA 19102
(215) 972-3400

36 SQUARE DE MEEUS, BTE 1
1040 BRUSSELS, BELGIUM
(02) 511 60 40

52 BEDFORD SQUARE
LONDON WC1B 3EX, ENGLAND
01. 631. 3383

1730 PENNSYLVANIA AVENUE, N.W
WASHINGTON, DC 20006
(202) 783-0200

800 NORTH THIRD STREET
HARRISBURG, PA 17102
(717) 233-7947

ROBERT A. COHEN

April 17, 1985

Mr. Michael Zinman
Earthworm Tractor Co. (Inc.)
495 Ashford Avenue
Ardsley, New York 10502

Re: <u>Raterpillar Tractor Co.</u>

Dear Michael:

Enclosed is your Closing Binder in connection with the sale of all of the outstanding shares of Raterpillar to Caterpillar.

Best regards.

Sincerely,

Robert

Robert A. Cohen

RAC:mc
Encl.

ACQUISITION OF ALL THE OUTSTANDING SHARES OF RATERPILLAR TRACTOR CO. BY CATERPILLAR TRACTOR CO.

INDEX

1. Share Acquisition Agreement (with Exhibits) by and between Michael Zinman and Caterpillar Tractor Co., dated April 9, 1985.

2. Power of Attorney executed by Caterpillar Tractor Co. in favor of R. Rennie Atterbury.

3. Closing Agenda.

4. Certificate of Seller certifying that all the representations, warranties and covenants of Seller contained in Article 4 of the Agreement are true and correct, and are fully complied with as at the Closing Date.

5. Opinion of Counsel for the Seller as to certain of the representations and warranties of the Seller.

6. Certificate from the Secretary of State of the State of New York concerning the good standing of the Company.

7. Certificate of Ronald M. Simons, independent certified accountant for the Company as to changes in financial conditions of the Company for the period from December 31, 1984 to the Closing Date.

8. Letter by Michael Zinman forgiving the Company's debt to him in the amount of $61.00.

9. Letter of Resignation and Release of Michael Zinman as President, Secretary and sole Director of the Company.

10. Stock Certificate No. 1, representing one (1) share (no par value) of the common stock of the Company, duly endorsed in bearer form by Seller, and receipt therefor issued by Buyer.

11. A receipt issued by Seller for a certified check issued by Buyer in the amount of U.S. $500.00 payable to the order of Michael Zinman.

12. Receipt issued by Buyer for corporate documents and seal delivered to Buyer by Seller.

SHARE ACQUISITION AGREEMENT

Agreement by and between Michael Zinman (herein called the "Seller"), and Caterpillar Tractor Co., a corporation organized and existing under the laws of California (herein called the "Buyer").

WHEREAS, Seller owns all of the issued and outstanding shares of Raterpillar Tractor Co. (herein called the "Company"), a corporation organized and existing under the laws of the State of New York, and wishes to sell such shares.

WHEREAS, Buyer wishes to buy one hundred percent (100%) of the shares of the Company.

NOW, THEREFORE, IN CONSIDERATION of the premises and the mutual covenants of the parties hereinafter set forth, it is agreed as follows:

Article 1. Definitions

(a) The "Closing" shall mean the closing of the transactions contemplated by this Agreement.

(b) The "Closing Date" shall mean the 11th day of April, 1985 or such other date as the parties may hereafter agree in writing.

(c) The "Share" shall mean the share identified and described in Article 2 hereof.

(d) The terms "material" or "materially" shall mean in an amount in excess of Fifty Dollars ($50.00).

Article 2. Sale and Purchase of Stock

Subject to the terms and conditions hereof, Seller agrees to sell to Buyer and, in reliance on the representations and warranties contained herein, Buyer agrees to purchase from Seller One (1) share of common stock of the Company, being one hundred percent (100%) of the issued and outstanding shares of capital stock of the Company, and Buyer shall issue a receipt therefor to Seller.

Article 3. Purchase Price and Payment for Share

3.1. Purchase Price. The total purchase price shall be Five Hundred Dollars ($500.00). Buyer shall pay the purchase price by delivering to Seller a certified check for Five Hundred Dollars ($500.00) and Seller shall issue a receipt therefor to Buyer.

Article 4. Seller's Representations and Warranties

Seller hereby makes the following representations, warranties and covenants to Buyer, knowing that each thereof is material to Buyer and is being relied on by Buyer in purchasing the Share.

4.1. Right to Sell and Authority. Seller has lawful authority to enter into this Agreement and to carry out the transactions contemplated hereby, including, without limitation, the sale and assignment of the Share to Buyer.

4.2. No Violation. Neither the execution nor the performance of this Agreement by Seller has resulted or will result in a violation of or conflict with any terms or provisions of the Articles of Incorporation or By-Laws of the Company or of any other agreement or instrument to which Seller or the Company, is a party or by which Seller or the Company are bound. Such execution and performance will not violate or conflict with any other restriction of any kind to which Seller or the Company, is subject.

4.3. Title to Sell. Seller is the lawful owner of and has good and marketable title, free and clear of any lien, claim, option, charge or encumbrance of any nature whatsoever, to the Share. By its purchase of the Share, Buyer will thereby become the owner and holder of the Share, free and clear of any lien, claim, option, charge or encumbrance of any nature whatsoever.

4.4. Organization, Standing and Authority of the Company. The Company is a corporation duly organized, validly existing and in good standing under the laws of the State of New York. The Company has full corporate power and lawful authority to carry on its business as now conducted and to own and operate its assets, properties and business.

4.5. Articles of Incorporation and By-Laws. Exhibit A attached hereto and made a part hereof is a copy of the Company's Articles of Incorporation (including all amendments thereto) certified by the Company's Secretary. Exhibit B attached hereto and made a part hereof is a copy of the Company's By-Laws (including all amendments thereto) certified by the Company's Secretary. Such copies are true, complete and correct as of the date of this Agreement. No change shall be made in such Articles of Incorporation or By-Laws through the Closing Date.

4.6. Directors and Officers. Exhibit C attached hereto and made a part hereof is a true, complete and correct list of all of the directors and officers of the Company as of the date hereof, together with information as to any existing employment contracts or agreements and the expiration of their terms of office or employment where finite.

4.7. Qualification to do Business. The State of New York is the only jurisdiction in which the Company is qualified to carry on business.

4.8. Capital Stock of the Company. As of the date hereof, Seller is the sole holder of all of the issued capital stock of the Company. The Share is validly issued, fully paid and non-assessible. There are no existing options, calls or commitments of any character whatsoever, or agreements to grant the same, relating to shares of the Company's capital stock, nor are there outstanding any securities convertible into or exercisable for such shares.

4.9. Subsidiaries of the Company. The Company does not own directly or indirectly any voting securities of any corporation or association. No company has been a subsidiary of the Company, and there is no company having any liability for which the Company is or will be legally responsible.

4.10. Financial Statements. Exhibit D attached hereto and made a part hereof contains copies of the balance sheet and statement of profit and loss of the Company as at December 31, 1984, certified by the Company's independent accounting firm. Such balance sheet and statement of profit and loss are correct and complete and correctly present, in accordance with generally accepted accounting principles, the financial condition of the Company at that date and the results of operations of the Company throughout the periods indicated. They have been prepared in accordance with generally accepted accounting principles applied on a consistent basis.

4.11. Undisclosed Liabilities. Neither Seller nor the Company know, or has any reasonable grounds to know, of any basis for the assertion against the Company of any liability of any material nature or any material amount. If during the period from the date hereof up to and including the Closing Date, Seller or the Company obtains knowledge or has reasonable grounds to know of any such liability, Seller will immediately notify Buyer of such liability, but such notification will not prejudice any rights of Buyer under this Agreement.

4.12. Absence of Certain Changes. Since December 31, 1984, there has not been (i) any change in the Company's financial condition, assets, liabilities or business, other than changes in the ordinary course of business which have not been materially adverse, individually or in the aggregate; (ii) any damage, destruction, or loss, whether or not covered by insurance, materially and adversely affecting the Company's assets individually or in the aggregate; (iii) any increase in the compensation payable or to be payable by the Company to any of its officers, employees or agents, or any bonus payment or arrangement made to or with any of them; (iv) any indebtedness incurred by the Company for borrowed money or any commitment to borrow money entered into by the Company; (v) any substantial revision or renegotiation of any contract with any party; or (vi) any other event or condition of any character, materially adversely affecting the financial position or operation of the business of the Company.

4.13. Rendering or Obtaining of Services. Since December 31, 1984, the Company has not rendered or obtained any services or entered into any transaction which has involved or will involve payment by the Company of any material amount or which will extend beyond the Closing Date, or which otherwise has affected or will affect its business. The Company will not render or obtain any service or enter into any contract during the period from the date hereof through the Closing Date.

4.14. Real Properties. The Company does not own, of record or beneficially, or lease any real property. During the period from the date hereof through the Closing Date, the Company will take no action which would render the representations and warranties contained in this Article 4.14 untrue or incorrect in any respect at and as of the Closing Date.

4.15. Title to Personal Property. As of the date hereof, the Company has good and marketable title to all its personal property shown in the books of the Company or used by it in its business, free and clear of all encumbrances, options, liens, charges or claims of any nature whatsoever, including any mortgage, pledge, conditional sale or other title retention agreement. Exhibit E attached hereto and made part hereof is an accurate and complete list as of the date hereof, with summary descriptions, of all personal property now owned or leased by the Company. From the date hereof through the Closing Date, the Company will not purchase, lease from another or otherwise acquire any personal property, or sell, lease to another or otherwise dispose of any personal property, or create any encumbrances, options, liens, charges or claims with respect thereto. During the period from the date hereof through the Closing Date, the Company will not take any action which would render the representations and warranties contained in this Article 4.15 untrue and incorrect in any respect at and as of the Closing Date.

4.16. Accounts and Notes Receivable. Any accounts and notes receivable on the books of the Company are subject to no material counterclaims or setoffs and are good and collectible at the aggregate recorded amounts thereof as shown on the books of the Company.

4.17. Tax Matters. The provision made for taxes on the Company's December 31, 1984 balance sheet is sufficient for the payment of all accrued and unpaid taxes of any nature whatsoever of the Company, whether or not disputed, for the calendar year ended on December 31, 1984 and for all periods prior thereto, including, without limitation, all income, sales and other taxes. There are no agreements by the Company for the extension of time for the assessment or payment of any tax, and all taxes due and payable by the Company on or before the date hereof have been paid.

4.18. Litigation. There is no litigation, arbitration, labor suit, governmental investigation or other legal or administrative proceeding whether pending or threatened or for which there is any basis known to Seller or the Company, against or relating to the Company or its business or properties or against or relating to the Company's directors, officers, employees or stockholders, in reference to actions taken by them in such capacity, or relating to the transactions contemplated by this Agreement which would have a material adverse effect on the Company's operations or assets or which would impose any material liability on the Company. There are no decrees, injunctions or orders of any court, governmental department or agency outstanding against the Company.

4.19. Compliance with Laws and Regulations. The Company is not in default with respect to, or charged with or threatened with a charge of violation of, or under investigation with respect to a possible violation of, any provision of any law, decree, administrative ruling or regulation of the United States or of any state thereof or any agreement with, or undertaking to, any administrative entity of the United States or any state thereof which relates to any aspect of the Company's business. The Company is in compliance with all governing laws or regulations applicable to it, the non-compliance with which would impose a material liability on or have a material adverse effect on the Company's operations or assets. All licenses, permits, orders or approvals necessary to the operations of the Company have been obtained by the Company and are in effect as of the date hereof.

4.20. Borrowings. The Company has no borrowings outstanding The Company is not liable in any manner in respect of any indebtedness (including directly or indirectly lending credit to, or guaranteeing any debts or obligations of, any other person, firm or corporation, whether by agreement to purchase or repurchase, by similar agreement or otherwise) for money borrowed or for money raised under acceptances or acceptance credits in relation to bills of exchange.

4.21. Loans. No person, firm or corporation is liable to the Company for any indebtedness (including directly or indirectly lending credit to, or guaranteeing any debts or obligations of, any other person, firm or corporation, whether by agreement to purchase or repurchase, by similar agreement or otherwise) for money borrowed. During the period from the date hereof through the Closing Date the Company will not lend any money.

4.22. Contracts and Commitments. As of the date hereof, the Company has not entered into any contract or commitment. During the period from the date hereof through the Closing Date, the Company will not enter into any contracts or commitments.

4.23. Industrial Property Rights. As of the date hereof, the Company does not possess nor does it have pending any application for United States or foreign patents, registered trademarks or trade names, unregistered trademarks or trade names, or copyrights or copyright registrations. Further the Company has never granted or been granted any license, or other agreement, relating in whole or in part to any industrial property rights, or to inventions, disclosures, improvements, processes, formulae, trade secrets, ideas or other know-how, or labels, whether or not patented or registered and whether owned by the Company or otherwise.

4.24. Insurance. As of the date hereof, the Company has no policies of insurance in force with respect to the Company, including, without limitation, those covering its employees, equipment, furniture and operations.

4.25. Salaries. Other than the officers named in Exhibit C attached hereto, the Company has as of the date hereof no employees. During the period from the date hereof through the Closing Date the Company shall take no action which would render the representations and warranties set forth in this Article 4.25 untrue or incorrect in any respect at and as of the Closing Date.

4.26. Employee Benefits. The Company has no employee welfare, pension and retirement plans and arrangements, including bonuses, percentage compensation, service awards and other like plans or benefits for officers or employers. There are no pensioned employees/of the Company. No bonus, percentage compensation, service award or other like benefit has been granted, made or accrued to or to the credit of any of the officers, employees or agents of the Company. During the period from the date hereof through the Closing Date, the Company shall take no action which would render the representations and warranties set forth in this Article 4.26 untrue or incorrect in any respect at and as of the Closing Date.

4.27. Bank Accounts. As of the date hereof, the Company has no bank account of any kind or safe deposit box. No change shall be made in such banking and safe deposit arrangements during the period from the date hereof through the Closing Date.

4.28. Powers of Attorney. As of the date hereof, the Company has never granted a power of attorney to any person or entity. No power of attorney will be granted by the Company during the period from the date hereof through the Closing Date.

4.29. Interest of Persons having a Special Relationship to the Company. For purposes of this Article 4.29 the term "persons having a special relationship to the Company" shall mean (i) all of the present officers, directors and stockholders (including the Seller) of the Company; (ii) all members of the family of any officer, director or stockholder (consisting of the grandparents of such officer, director or stockholder, all descendants of such grandparents (including by adoption) and all spouses of such descendants (including any spouse of any officer, director or stockholder); and (iii) any legal entity, including without limitation any corporation, partnership, joint venture or trust, in which any one or more of such persons having a special relationship to the Company shall have in the aggregate, directly or indirectly, an interest of 10% or more in either the capital, the business or the income of such legal entity. No person having a special relationship to the Company, as defined above, has:

(a) any right or title to, or interest in, the Company, other than the ownership by the Seller of his Shares;

(b) any right, title or interest in any property, real or personal, tangible or intangible, which is owned by, or is in the possession of the Company; or

(c) any claim or contractual right of any nature whatsoever, whether accrued or contingent, against the Company, or is owed any amount by the Company.

4.30. Conduct of Business Prior to the Closing. In addition to the covenants given heretofore with respect to the Company's actions on certain specific matters during the period from the date hereof through the Closing Date, Seller further covenants that during such period the Company will not take any action, which will render any of the representations and warranties contained in this Article 4.30 untrue or incomplete in any respect at and as of the Closing Date. Without limiting the generality of the foregoing, Seller further specifically covenants that during such period, the Company will maintain its present condition.

4.31. No Misleading Statement or Material Omission. The representations and warranties of Seller set forth in this Agreement and in the Exhibits attached hereto do not contain any untrue statements of fact and do not omit the disclosure, whether or not called for hereby, of any fact which is known to, or which should be known to, Seller, and which would have the effect of making any statement contained therein misleading or which might reasonably be expected to affect adversely and materially the Company.

Article 5.
Access to Information Concerning the Company Prior to the Closing. Seller will cause the Company to give to Buyer and Buyer's counsel, accountants and other representatives full access, during normal business hours throughout the period prior to the Closing, to all the properties, books, contracts, commitments and records of the Company and will cause the Company to furnish Buyer during such period with all such information concerning the affairs of the Company as Buyer shall reasonably request, including without limitation, existing interim financial and historical information about the Company and true and complete copies, certified by the Company's Secretary, of any of the documents referred to in Article 4 hereof. It is understood by Seller and Buyer that Buyer has no duty or obligation to make any investigation and such investigations as Buyer may make shall not relieve Seller in any manner whatsoever from any of the representations and warranties made hereunder nor shall such investigations be deemed to alter, amend, modify, limit or restrict the scope, amount or meaning of such representations and warranties.

Article 6. Conditions Precedent to Buyer's Obligations Hereunder

All obligations of Buyer under this Agreement are subject to the fulfillment to the satisfaction of Buyer (or waiver in writing by Buyer), prior to or at the Closing, of each of the following conditions:

6.1. Representations, etc., as of the Date of this Agreement. Buyer shall not have discovered any material error, misstatement, or omission in the representations and warranties made by Seller in or under this Agreement.

6.2. Representations, etc., as of Closing Date. Seller's representations and warranties made in or under this Agreement shall be deemed to have been made again at and as of the time of Closing, and shall then be true and correct in all respects; Seller shall have performed and complied with all agreements, covenants and conditions required by this Agreement to be performed and complied with by them prior to or at the Closing; and Buyer shall have been furnished with certificates of Seller, dated the Closing Date, certifying in such detail as Buyer may reasonably request to the fulfillment thereof.

6.4. Accountants' Letter. Buyer shall have received from the Company's independent certified public accountants a letter dated not earlier than five days prior to the Closing Date, in form and substance satisfactory to Buyer, to the effect that, on the basis of a limited review, but not an audit, of the books and records of the Company, including minutes of meetings of shareholders and directors and committees thereof, and conversations with the principal officers of the Company, nothing has come to the attention of said accountants which would indicate that since December 31, 1984 there has been any materially adverse change in the consolidated financial condition of the Company as reflected in the financial statements of the Company as of December 31, 1984.

6.5. Opinion of Counsel for Seller. Buyer shall have been furnished with an opinion, dated the Closing Date, of Dechert Price & Rhoads, counsel for Seller and the Company, to the effect that the statements set forth in Articles 4.1, 4.2 and 4.4 through 4.9 hereof are true and correct as of the Closing Date.

Article 7. Closing

7.1. Time and Place. The Closing shall take place at the law offices of Dechert, Price & Rhoads, 477 Madison Avenue, New York, New York at 4:00 P.M. on the Closing Date, or at such other place or at such other time or on such other date as Seller and Buyer shall agree.

7.2. Conveyance of Shares. At the Closing, Seller will deliver to Buyer good and marketable title to the Shares, free and clear of all liens, claims, options, charges and encumbrances of any nature whatsoever, by delivering to Buyer the Shares, appropriately endorsed in bearer form and delivered to Buyer and Buyer shall issue a receipt therefor to Seller. Seller shall have paid as of the Closing Date all requisite documentary transfer taxes in full.

7.3. Resignations. At or prior to the Closing, Seller shall deliver to Buyer the resignations of all of the Company's officers, directors, agents, employees and representatives, which resignations shall release, acquit, remise and forever waive and discharge the Company and Buyer from any and all claims, demands, actions, causes of action, damages and costs of every kind and nature in any way relating to their employment by the Company.

7.4. Payment of Consideration by Buyer. Subject to the terms and conditions of this Agreement in full consideration for the sale, conveyance, transfer and delivery of the Shares, Buyer shall pay Seller at the Closing an amount equal to Five Hundred Dollars ($500.00), by delivering to Seller a certified check for that amount, and Seller shall issue a receipt to Buyer.

Article 8. Rescission

8.1. Non-fulfillment of Conditions Precedent. If, despite the best efforts of the parties, the conditions precedent to Closing set forth in Article 6 hereof are not fulfilled by the Closing Date specified in Article 7.1 hereof, then either party may, without judicial resolution and without incurring any liability to the other party, rescind this Agreement by serving written notice upon the other party.

8.2. Breach of Agreement. In the event of a material breach on or before the Closing Date by Seller of any of the representations, warranties or obligations set forth in this Agreement, which has not been cured within thirty (30) days after notice thereof to Seller by Buyer, then Buyer may, at its sole discretion, without judicial resolution, without in any way limiting or modifying any of the terms and conditions of this Agreement, or any of the remedies to which it would otherwise be entitled, and without incurring any liability of any kind whatsoever to Seller, rescind this Agreement by serving written notice upon Seller, it being understood and agreed that the aforesaid thirty (30) day period after notice of a breach shall in no event extend beyond five business days after the Closing Date specified in Section 7.1 hereof and that such Date shall be subject to a single extension for purposes of this Section 8.2.

Article 9. Indemnities

9.1. Without in any way limiting or modifying any of the terms and conditions of this Agreement, or any of the remedies to which Buyer would otherwise be entitled, Seller hereby agrees to hold Buyer harmless from, and to indemnify Buyer against any loss, liability, damage or expense, including court costs and reasonable legal fees, suffered or paid by Buyer arising, directly or indirectly, (i) out of any breach of Seller's representations and warranties set forth in this Agreement or (ii) out of any claim, demand, penalty, fine or tax, except such corporate franchise tax which may be attributed to the operation of the Company subsequent to December 31, 1984, asserted or imposed against Buyer as transferee of the Shares by any governmental entity or agency or by any other party, with respect to the conduct of business by the Company prior to the Closing Date.

9.2 In the event that Buyer shall be threatened with any loss, liability, damage or expense of the type referred to in Article 9.1 above, the Buyer shall give written notice thereof, and Seller shall, within thirty (30) days following receipt of said notice, pay, settle, discharge or make Buyer whole with respect to the same or hold Buyer harmless therefrom.

Article 10. Non-Competition

Seller agrees that for a period of ten (10) years from the Closing Date neither Seller nor any person, firm or corporation of which Seller owns, of record or beneficially, fifty percent (50%) or more of the equity interest, shall register or use, in any way, directly or indirectly, any name, word or mark, confusingly similar to any name, trademark or service mark, registered or unregistered, owned, hereafter acquired or used by Buyer or any of its affiliated companies in any country.

Article 11. Finders and Brokers

Seller and Buyer agree that no person, firm or corporation is entitled to receive from either of them any finders or brokerage commissions arising out of the transactions contemplated by this Agreement, and if any claim is made by any finder or broker claiming to have acted on behalf of either party, such party shall indemnify the other party for any amount for which such other party may be held liable with respect thereto, and all expenses reasonably incurred by such other party in defending against such claim.

Article 12. Notices, etc.

12.1. All notices requests, consents, demands, waivers or other communications hereunder shall be in writing and shall be sent by hand delivery to the persons specified below or by prepaid first class registered airmail to the following addresses:

If to Sellers to:
Michael Zinman
495 Ashford Avenue
Ardsley, New York 10502

If to Buyer:
Caterpillar Tractor Co.
100 N. E. Adams Street
Peoria, Illinois 61629

Attention: Secretary

12.2. The parties may at any time designate, by like notice, other addresses to which notices and other communications should be sent.

Article 13. Expenses and Fees

All expense involved in the preparation, authorization and consummation of this Agreement and the Exhibits hereto, including, without limitation, all fees and expenses of agents, representatives, counsel and accountants in connection therewith, shall be borne solely by the party which shall have incurred the same, and the other party shall have no liability in respect thereof.

Article 14. Assignability

The rights and liabilities of the parties to this Agreement shall bind and inure to their respective successors and assigns; provided, however, that this Agreement and all rights and duties hereunder shall not be assignable or delegable without the express written consent of the other party.

Article 15. Waivers

There shall be no waiver of any term, provision or condition of this Agreement unless said waiver is evidenced in a written document duly signed on behalf of the waiving party. No such waiver shall be deemed to be or construed as a further continuing waiver of any such term, provision or condition unless the writing which evidences such waiver states to the contrary.

Article 16.
Nature of Certificates, Disclosure and Survival of Representations

All statements contained in any certificate or other instrument delivered by or on behalf of Seller pursuant hereto or in connection with the transactions contemplated hereby shall be deemed representations and warranties hereunder. No representation or warranty made by Seller in this Agreement or made pursuant hereto, or in connection with the transaction contemplated hereby, contains or will contain any untrue statement of any fact, or omits or will omit to state any fact necessary to make the statements contained therein not misleading. All representations and warranties contained in this Agreement or made pursuant thereto shall survive the Closing and any investigation of fact.

Article 17. Article and Other Headings

The article and other headings contained in this Agreement are for reference purposes only and shall not affect in any way the meaning or interpretation of this Agreement.

Article 18. Entire Agreement and Amendments

This Agreement, with its Exhibits, constitutes the entire agreement between the parties with respect to the matters covered herein, and supersedes and cancels any prior writings and understandings, if any, with respect to the matters covered herein. This Agreement may be amended or modified only by a written instrument executed by the parties hereto.

Article 19. Governing Law

This Agreement shall be governed by the laws of the State of Illinois.

IN WITNESS WHEREOF, the parties have executed this Share Acquisition Agreement in two (2) counterparts.

Michael Zinman

Date 8 April 1985

Caterpillar Tractor Co.

By R. Rennie Atterbury

Title Associate Counsel

Date April 9, 1985

RATERPILLAR TRACTOR COMPANY, INC.

CERTIFICATE OF THE SECRETARY

I, Michael Zinman, Secretary of RATERPILLAR TRACTOR CO., INC. (the "Company") do hereby certify that attached hereto as Exhibit A is a true, correct and complete copy of the Certificate of Incorporation of the Company, filed on January 26, 1984 as certified by the Secretary of State of the State of New York on July 24, 1984. Such Certificate of Incorporation is in full force and effect as of the date hereof, no amendments having been made to such Certificate of Incorporation since January 26, 1984.

IN WITNESS WHEREOF, I have hereunto signed my name and affixed the seal of the Company this 8th day of April, 1985.

[CORPORATE SEAL]

Michael Zinman
Secretary

State of New York } ss.:
Department of State

43937

I hereby certify that I have compared the annexed copy with the original document filed by the Department of State and that the same is a correct transcript of said original.

Witness my hand and seal of the Department of State on JUL 24 1984.

Secretary of State

380607-004(12/82)

CT

CERTIFICATE OF INCORPORATION
OF
RATERPILLAR TRACTOR CO., INC.

UNDER SECTION 402 OF THE
BUSINESS CORPORATION LAW

OK upon cancel of Ken. Paper

DECHERT PRICE & RHOADS
477 MADISON AVENUE
NEW YORK, NEW YORK 10022

STATE OF NEW YORK
DEPARTMENT OF STATE
FILED JAN 26 1984

BILLED

B062753

FILED

Certificate of Incorporation
of
Raterpillar Tractor Co., Inc.

(Under Section 402 of the Business Corporation Law)

The undersigned, being a natural person of at least 18 years of age, and acting as the incorporator of the corporation, hereby being formed under the Business Corporation Law of the State of New York, certifies:

First: The name of the corporation shall be Raterpillar Tractor Co., Inc.

Second: The corporation is formed for the following purposes:

To engage in any lawful act or activity for which a corporation may be organized under the General Corporation Law of New York, provided that the corporation is not formed to engage in any act or activity which requires the consent or approval of any state official, department, board, agency or other governmental body.

Third: The city and county within the State of New York in which the office of the corporation is to be located, are as follows: Ardsley, County of Westchester.

Fourth: The aggregate number of shares which the corporation shall have authority to issue is two hundred (200) shares, all of which are no par, common and all of which are of the same class.

Fifth: The Secretary of State is designated as the agent of the corporation upon whom process in any action or proceeding against the corporation may be served. The address to which the Secretary of State shall mail a copy of any process against the corporation served upon him is c/o Dechert Price & Rhoads, 477 Madison Avenue, New York, New York 10022.

IN WITNESS WHEREOF, the undersigned has executed and acknowledged this Certificate this 24th day of January, 1984.

Laura L. Inman

Laura L. Inman, Incorporator
477 Madison Avenue
New York, New York 10022

STATE OF NEW YORK)
: ss:
COUNTY OF NEW YORK)

On the date hereinafter set forth before me personally came Laura L. Inman to me known to be the person described in and who executed the foregoing certificate and she acknowledged to me that she executed the same.

Dated: January 24, 1984

RATERPILLAR TRACTOR COMPANY, INC.

CERTIFICATE OF THE SECRETARY

I, Michael Zinman, Secretary of RATERPILLAR TRACTOR CO., INC. (the "Company") do hereby certify that attached hereto as Exhibit A is a true, correct and complete copy of the By-Laws of the Company, as duly adopted at the organizational meeting of the Company on February 15, 1984. Such By-Laws are in full force and effect as of the date hereof, no amendments having been made to such By-Laws since February 15, 1984.

IN WITNESS WHEREOF, I have hereunto signed my name and affixed the seal of the Company this 8th day of April, 1985.

[CORPORATE SEAL]

Michael Zinman
Secretary

BY-LAWS

of

RATERPILLAR TRACTOR CO., INC.

ARTICLE I - OFFICES

The principal office of the corporation shall be in the town of Ardsley County of Westchester State of New York. The corporation may also have offices at such other places within or without the State of New York as the board may from time to time determine or the business of the corporation may require.

ARTICLE II - SHAREHOLDERS

1. PLACE OF MEETINGS.

Meetings of shareholders shall be held at the principal office of the corporation or at such place within or without the State of New York as the board shall authorize.

2. ANNUAL MEETING.

The annual meeting of the shareholders shall be held on the 30th day of March at 2 P.M. in each year if not a legal holiday, and, if a legal holiday, then on the next business day following at the same hour, when the shareholders shall elect a board and transact such other business as may properly come before the meeting.

3. SPECIAL MEETINGS.

Special meetings of the shareholders may be called by the board or by the president and shall be called by the president or the secretary at the request in writing of a majority of the board or at the request in writing by shareholders owning a majority in amount of the shares issued and outstanding. Such request shall state the purpose or purposes of the proposed meeting. Business transacted at a special meeting shall be confined to the purposes stated in the notice.

4. FIXING RECORD DATE.

For the purpose of determining the shareholders entitled to notice of or to vote at any meeting of shareholders or any adjournment thereof, or to express consent to or dissent from any proposal without a meeting, or for the purpose of determining shareholders entitled to receive payment of any dividend or the allotment of any rights, or for the purpose of any other action, the board shall fix, in advance, a date as the record date for any such determination of shareholders. Such date shall not be more than fifty nor less than ten days before the date of such meeting, nor more than fifty days prior to any other action. If no record date is fixed it shall be determined in accordance with the provisions of law.

5. NOTICE OF MEETINGS OF SHAREHOLDERS.

Written notice of each meeting of shareholders shall state the purpose or purposes for which the meeting is called, the place, date and hour of the meeting and unless it is the annual meeting, shall indicate that it is being issued by or at the direction of the person or persons calling the meeting. Notice shall be given either personally or by mail to each shareholder entitled to vote at such meeting, not less than ten nor more than fifty days before the date of the meeting. If action is proposed to be taken that might entitle shareholders to payment for their shares, the notice shall include a statement of that purpose and to that effect. If mailed, the notice is given when deposited in the United States mail, with postage thereon prepaid, directed to the shareholder at his address as it appears on the record of shareholders, or, if he shall have filed with the secretary a written request that notices to him be mailed to some other address, then directed to him at such other address.

6. WAIVERS.

Notice of meeting need not be given to any shareholder who signs a waiver of notice, in person or by proxy, whether before or after the meeting. The attendance of any shareholder at a meeting, in person or by proxy, without protesting prior to the conclusion of the meeting the lack of notice of such meeting, shall constitute a waiver of notice by him.

7. QUORUM OF SHAREHOLDERS.

Unless the certificate of incorporation provides otherwise, the holders of a majority of the shares entitled to vote thereat shall constitute a quorum at a meeting of shareholders for the transaction of any business, provided that when a specified item of business is required to be voted on by a class or classes, the holders of a majority of the shares of such class or classes shall constitute a quorum for the transaction of such specified item of business.

When a quorum is once present to organize a meeting, it is not broken by the subsequent withdrawal of any shareholders.

The shareholders present may adjourn the meeting despite the absence of a quorum.

8. PROXIES.

Every shareholder entitled to vote at a meeting of shareholders or to express consent or dissent without a meeting may authorize another person or persons to act for him by proxy.

Every proxy must be signed by the shareholder or his attorney-in-fact. No proxy shall be valid after expiration of eleven months from the date thereof unless otherwise provided in the proxy. Every proxy shall be revocable at the pleasure of the shareholder executing it, except as otherwise provided by law.

9. QUALIFICATION OF VOTERS.

Every shareholder of record shall be entitled at every meeting of shareholders to one vote for every share standing in his name on the record of shareholders, unless otherwise provided in the certificate of incorporation.

10. VOTE OF SHAREHOLDERS.

Except as otherwise required by statute or by the certificate of incorporation;

(a) directors shall be elected by a plurality of the votes cast at a meeting of shareholders by the holders of shares entitled to vote in the election;

(b) all other corporate action shall be authorized by a majority of the votes cast.

11. WRITTEN CONSENT OF SHAREHOLDERS.

Any action that may be taken by vote may be taken without a meeting on written consent, setting forth the action so taken, signed by the holders of all the outstanding shares entitled to vote thereon or signed by such lesser number of holders as may be provided for in the certificate of incorporation.

ARTICLE III - DIRECTORS

1. BOARD OF DIRECTORS.

Subject to any provision in the certificate of incorporation the business of the corporation shall be managed by its board of directors, each of whom shall be at least 18 years of age and need not be shareholders.

2. NUMBER OF DIRECTORS.

The number of directors shall be three
When all of the shares are owned by less than three shareholders, the number of directors may be less than three but not less than the number of shareholders.

3. ELECTION AND TERM OF DIRECTORS.

At each annual meeting of shareholders, the shareholders shall elect directors to hold office until the next annual meeting. Each director shall hold office until the expiration of the term for which he is elected and until his successor has been elected and qualified, or until his prior resignation or removal.

4. NEWLY CREATED DIRECTORSHIPS AND VACANCIES.

Newly created directorships resulting from an increase in the number of directors and vacancies occurring in the board for any reason except the removal of directors without cause may be filled by a vote of a majority of the directors then in office, although less than a quorum exists, unless otherwise provided in the certificate of incorporation. Vacancies occurring by reason of the removal of directors without cause shall be filled by vote of the shareholders unless otherwise provided in the certificate of incorporation. A director elected to fill a vacancy caused by resignation, death or removal shall be elected to hold office for the unexpired term of his predecessor.

5. REMOVAL OF DIRECTORS.

Any or all of the directors may be removed for cause by vote of the shareholders or by action of the board. Directors may be removed without cause only by vote of the shareholders.

6. RESIGNATION.

A director may resign at any time by giving written notice to the board, the president or the secretary of the corporation. Unless otherwise specified in the notice, the resignation shall take effect upon receipt thereof by the board or such officer, and the acceptance of the resignation shall not be necessary to make it effective.

7. QUORUM OF DIRECTORS.

Unless otherwise provided in the certificate of incorporation, a majority of the entire board shall constitute a quorum for the transaction of business or of any specified item of business.

8. ACTION OF THE BOARD.

Unless otherwise required by law, the vote of a majority of the directors present at the time of the vote, if a quorum is present at such time, shall be the act of the board. Each director present shall have one vote regardless of the number of shares, if any, which he may hold.

9. PLACE AND TIME OF BOARD MEETINGS.

The board may hold its meetings at the office of the corporation or at such other places, either within or without the State of New York, as it may from time to time determine.

10. REGULAR ANNUAL MEETING.

A regular annual meeting of the board shall be held immediately following the annual meeting of shareholders at the place of such annual meeting of shareholders.

11. NOTICE OF MEETINGS OF THE BOARD, ADJOURNMENT.

(a) Regular meetings of the board may be held without notice at such time and place as it shall from time to time determine. Special meetings of the board shall be held upon notice to the directors and may be called by the president upon three days notice to each director either personally or by mail or by wire; special meetings shall be called by the president or by the secretary in a like manner on written request of two directors. Notice of a meeting need not be given to any director who submits a waiver of notice whether before or after the meeting or who attends the meeting without protesting prior thereto or at its commencement, the lack of notice to him.

(b) A majority of the directors present, whether or not a quorum is present, may adjourn any meeting to another time and place. Notice of the adjournment shall be given all directors who were absent at the time of the adjournment and, unless such time and place are announced at the meeting, to the other directors.

12. CHAIRMAN.

At all meetings of the board the president, or in his absence, a chairman chosen by the board shall preside.

13. EXECUTIVE AND OTHER COMMITTEES.

The board, by resolution adopted by a majority of the entire board, may designate from among its members an executive committee and other committees, each consisting of three or more directors. Each such committee shall serve at the pleasure of the board.

14. COMPENSATION.

No compensation shall be paid to directors, as such, for their services, but by resolution of the board a fixed sum and expenses for actual attendance, at each regular or special meeting of the board may be authorized. Nothing herein contained shall be construed to preclude any director from serving the corporation in any other capacity and receiving compensation therefor.

ARTICLE IV - OFFICERS

1. OFFICES, ELECTION, TERM.

(a) Unless otherwise provided for in the certificate of incorporation, the board may elect or appoint a president, one or more vice-presidents, a secretary and a treasurer, and such other officers as it may determine, who shall have such duties, powers and functions as hereinafter provided.

(b) All officers shall be elected or appointed to hold office until the meeting of the board following the annual meeting of shareholders.

(c) Each officer shall hold office for the term for which he is elected or appointed and until his successor has been elected or appointed and qualified.

2. REMOVAL, RESIGNATION, SALARY, ETC.

(a) Any officer elected or appointed by the board may be removed by the board with or without cause.

(b) In the event of the death, resignation or removal of an officer, the board in its discretion may elect or appoint a successor to fill the unexpired term.

(c) Any two or more offices may be held by the same person, except the offices of president and secretary. When all of the issued and outstanding stock of the corporation is owned by one person, such person may hold all or any combination of offices.

(d) The salaries of all officers shall be fixed by the board.

(e) The directors may require any officer to give security for the faithful performance of his duties.

3. PRESIDENT.

The president shall be the chief executive officer of the corporation; he shall preside at all meetings of the shareholders and of the board; he shall have the management of the business of the corporation and shall see that all orders and resolutions of the board are carried into effect.

4. VICE-PRESIDENTS.

During the absence or disability of the president, the vice-president, or if there are more than one, the executive vice-president, shall have all the powers and functions of the president. Each vice-president shall perform such other duties as the board shall prescribe.

5. SECRETARY.

The secretary shall:

(a) attend all meetings of the board and of the shareholders;

(b) record all votes and minutes of all proceedings in a book to be kept for that purpose;

(c) givé or cause to be given notice of all meetings of shareholders and of special meetings of the board;

(d) keep in safe custody the seal of the corporation and affix it to any instrument when authorized by the board;

(e) when required, prepare or cause to be prepared and available at each meeting of shareholders a certified list in alphabetical order of the names of shareholders entitled to vote thereat, indicating the number of shares of each respective class held by each;

(f) keep all the documents and records of the corporation as required by law or otherwise in a proper and safe manner.

(g) perform such other duties as may be prescribed by the board.

6. ASSISTANT-SECRETARIES.

During the absence or disability of the secretary, the assistant-secretary, or if there are more than one, the one so designated by the secretary or by the board, shall have all the powers and functions of the secretary.

7. TREASURER.

The treasurer shall:

(a) have the custody of the corporate funds and securities;

(b) keep full and accurate accounts of receipts and disbursements in the corporate books;

(c) deposit all money and other valuables in the name and to the credit of the corporation in such depositories as may be designated by the board;

(d) disburse the funds of the corporation as may be ordered or authorized by the board and preserve proper vouchers for such disbursements;

(e) render to the president and board at the regular meetings of the board, or whenever they require it, an account of all his transactions as treasurer and of the financial condition of the corporation;

(f) render a full financial report at the annual meeting of the shareholders if so requested;

(g) be furnished by all corporate officers and agents at his request, with such reports and statements as he may require as to all financial transactions of the corporation;

(h) perform such other duties as are given to him by these by-laws or as from time to time are assigned to him by the board or the president.

8. ASSISTANT-TREASURER.

During the absence or disability of the treasurer, the assistant-treasurer, or if there are more than one, the one so designated by the secretary or by the board, shall have all the powers and functions of the treasurer.

9. SURETIES AND BONDS.

In case the board shall so require, any officer or agent of the corporation shall execute to the corporation a bond in such sum and with such surety or sureties as the board may direct, conditioned upon the faithful performance of his duties to the corporation and including responsibility for negligence and for the accounting for all property, funds or securities of the corporation which may come into his hands.

ARTICLE V - CERTIFICATES FOR SHARES

1. CERTIFICATES.

The shares of the corporation shall be represented by certificates. They shall be numbered and entered in the books of the corporation as they are issued. They shall exhibit the holder's name and the number of shares and shall be signed by the president or a vice-president and the treasurer or the secretary and shall bear the corporate seal.

2. LOST OR DESTROYED CERTIFICATES.

The board may direct a new certificate or certificates to be issued in place of any certificate or certificates theretofore issued by the corporation, alleged to have been lost or destroyed, upon the making of an affidavit of that fact by the person claiming the certificate to be lost or destroyed. When authorizing such issue of a new certificate or certificates, the board may, in its discretion and as a condition precedent to the issuance thereof, require the owner of such lost or destroyed certificate or certificates, or his legal representative, to advertise the same in such manner as it shall require and/or give the corporation a bond in such sum and with such surety or sureties as it may direct as indemnity against any claim that may be made against the corporation with respect to the certificate alleged to have been lost or destroyed.

3. TRANSFERS OF SHARES.

(a) Upon surrender to the corporation or the transfer agent of the corporation of a certificate for shares duly endorsed or accompanied by proper evidence of succession, assignment or authority to transfer, it shall be the duty of the corporation to issue a new certificate to the person entitled thereto, and cancel the old certificate; every such transfer shall be entered on the transfer book of the corporation which shall be kept at its principal office. No transfer shall be made within ten days next preceding the annual meeting of shareholders.

(b) The corporation shall be entitled to treat the holder of record of any share as the holder in fact thereof and, accordingly, shall not be bound to recognize any equitable or other claim to or interest in such share on the part of any other person whether or not it shall have express or other notice thereof, except as expressly provided by the laws of New York.

4. CLOSING TRANSFER BOOKS.

The board shall have the power to close the share transfer books of the corporation for a period of not more than ten days during the thirty day period immediately preceding (1) any shareholders' meeting, or (2) any date upon which shareholders shall be called upon to or have a right to take action without a meeting, or (3) any date fixed for the payment of a dividend or any other form of distribution, and only those shareholders of record at the time the transfer books are closed, shall be recognized as such for the purpose of (1) receiving notice of or voting at such meeting, or (2) allowing them to take appropriate action, or (3) entitling them to receive any dividend or other form of distribution.

ARTICLE VI - DIVIDENDS

Subject to the provisions of the certificate of incorporation and to applicable law, dividends on the outstanding shares of the corporation may be declared in such amounts and at such time or times as the board may determine. Before payment of any dividend, there may be set aside out of the net profits of the corporation available for dividends such sum or sums as the board from time to time in its absolute discretion deems proper as a reserve fund to meet contingencies, or for equalizing dividends, or for repairing or maintaining any property of the corporation, or for such other purpose as the board shall think conducive to the interests of the corporation, and the board may modify or abolish any such reserve.

ARTICLE VII - CORPORATE SEAL

The seal of the corporation shall be circular in form and bear the name of the corporation, the year of its organization and the words "Corporate Seal, New York." The seal may be used by causing it to be impressed directly on the instrument or writing to be sealed, or upon adhesive substance affixed thereto. The seal on the certificates for shares or on any corporate obligation for the payment of money may be a facsimile, engraved or printed.

ARTICLE VIII - EXECUTION OF INSTRUMENTS

All corporate instruments and documents shall be signed or countersigned, executed, verified or acknowledged by such officer or officers or other person or persons as the board may from time to time designate.

ARTICLE IX - FISCAL YEAR

The fiscal year shall begin the first day of January in each year.

ARTICLE X - REFERENCES TO CERTIFICATE OF INCORPORATION

Reference to the certificate of incorporation in these by-laws shall include all amendments thereto or changes thereof unless specifically excepted.

ARTICLE XI - BY-LAW CHANGES

AMENDMENT, REPEAL, ADOPTION, ELECTION OF DIRECTORS.

(a) Except as otherwise provided in the certificate of incorporation the by-laws may be amended, repealed or adopted by vote of the holders of the shares at the time entitled to vote in the election of any directors. By-laws may also be amended, repealed or adopted by the board but any by-law adopted by the board may be amended by the shareholders entitled to vote thereon as hereinabove provided.

(b) If any by-law regulating an impending election of directors is adopted, amended or repealed by the board, there shall be set forth in the notice of the next meeting of shareholders for the election of directors the by-law so adopted, amended or repealed, together with a concise statement of the changes made.

EXHIBIT C

As of the date of this Agreement, the following person holds the offices of the Company set out opposite his name:

Michael Zinman - President
Secretary
Director

These are the only offices of the Company which have been filled. There are no contracts or other agreements between the Company and the person named above relating to the terms of employment of the offices of the Company held by such person. No compensation has been paid to or is due such person for his services in the offices of the Company held by such person.

RATERPILLAR TRACTOR CO., INC.

FINANCIAL STATEMENTS

JANUARY 24, 1984 - DECEMBER 31, 1984

RONALD M. SIMONS
CERTIFIED PUBLIC ACCOUNTANT
9 DREXEL COURT
SPRING VALLEY, N. Y. 10977
(914)-352-6407

March 29, 1985

Board of Directors
Raterpillar Tractor Co., Inc.
Ardsley, New York 10502

I have examined the balance sheet of Raterpillar Tractor Co., Inc. at December 31, 1984 and the related statement of operations for the period January 24, 1984 through December 31, 1984. My examination was made in accordance with generally accepted auditing standards and, accordingly, included such tests of the accounting records and such other auditing procedures as I considered necessary in the circumstances.

In my opinion, the above mentioned financial statements present fairly the financial position of Raterpillar Tractor Co., Inc. at December 31, 1984 and the results of operations for the period then ended in conformity with generally accepted accounting principles applied on a consistent basis.

RATERPILLAR TRACTOR CO., INC.

STATEMENT OF OPERATIONS

JANUARY 24 - DECEMBER 31, 1984

REVENUES	- 0 -
EXPENSES:	
TAXES	293
OFFICE SUPPLIES	28
TOTAL	321
NET LOSS	(321)
RETAINED EARNINGS - BEGINNING	- 0 -
DEFICIT - DECEMBER 31, 1984	$ (321)

RATERPILLAR TRACTOR CO., INC.

BALANCE SHEET

DECEMBER 31, 1984

ASSETS	
CURRENT:	
CASH	$ 1
PREPAID SUPPLIES	52
TOTAL	$ 53
LIABILITIES AND CAPITAL DEFICIT	
CURRENT LIABILITIES:	
ACCRUED TAXES PAYABLE	$ 293
LOANS PAYABLE - OFFICER	61
TOTAL CURRENT LIABILITIES	354
CAPITAL DEFICIT:	
COMMON STOCK - NO PAR VALUE - 1 SHARE	20
DEFICIT	(321)
TOTAL CAPITAL DEFICIT	(301)
TOTAL	$ 53

EXHIBIT E

The following is a list of all of the personal property owned or leased by the Company as of the date hereof:

Quantity	Description	Value[1]
Two (2)[3]	Rats [2]	$6,98

1) "Value" is defined as cost to the Company in purchasing the item of personal property.

2) Rats are furry animals described in the American Heritage Dictionary as "long-tailed rodents resembling, but larger then, mice; especially one of the genus Rattus." Seller makes no representation or warranty as to the pedigree of the rats owned by the Company.

3) The quantity of this item of personal property is subject to change without any action on the part of Seller or the Company due to the death of existing or the birth of additional such items of personal property.

CATERPILLAR TRACTOR CO.

Peoria, Illinois 61629-7310

POWER OF ATTORNEY

KNOW ALL MEN BY THESE PRESENTS, That Caterpillar Tractor Co., a corporation duly organized and existing under the laws of the State of California, United States of America, and having its principal office at Peoria, Illinois, United States of America, does by these presents hereby nominate, R. Rennie Atterbury III its true and lawful attorney in fact, in its name, place and stead, and on its behalf, to execute a Share Acquisition Agreement between Caterpillar Tractor Co. and Raterpillar Tractor Co. for the purchase by Caterpillar Tractor Co. of One Hundred Percent (100%) of the issued and outstanding common stock of Raterpillar Tractor Co., and to do all other acts in connection therewith that may be necessary or appropriate, granting and giving unto said attorney full authority and power to do and perform any and all other acts necessary or incidental to the performance and execution of the powers herein expressly granted as fully to all intents and purposes as the grantor might or could do if personally present, with full power of substitution.

IN WITNESS WHEREOF, Caterpillar Tractor Co. has caused this Power of Attorney to be executed by its duly authorized representative as of this 8th day of April, 1985.

CATERPILLAR TRACTOR CO.

By Peter P. Donis
Executive Vice President

(S E A L)

ATTEST:

Assistant Secretary

p11

CLOSING AGENDA

Reference is here made to the Share Acquisition Agreement (the "Agreement") by and between Michael Zinman and Caterpillar Tractor Co., a California corporation, dated April 9, 1985. All defined terms used herein are those defined in the Agreement unless otherwise defined herein.

The Closing took place at the offices of Dechert Price & Rhoads located at 477 Madison Avenue, New York, New York, at 4:00 P.M. on the 11th day of April, 1985. Present were Michael Zinman, President, Secretary, sole Director and sole shareholder of the Company; R. Rennie Atterbury, Associate Counsel of Caterpillar Tractor Co.; and Robert Cohen and Karl Okamoto of Dechert Price & Rhoads, counsel for Seller. To effectuate the Closing, the following actions were simultaneously taken:

A. Seller delivered to Buyer:

1. Certificate of Seller certifying that all the representations, warranties and covenants of Seller contained in Article 4 of the Agreement are true and correct, and are fully complied with as at the Closing Date.

2. Opinion of Counsel for the Seller as to certain of the representations and warranties of the Seller.

3. Certificate with confirmatory telegram from the Secretary of State of the State of New York concerning the good standing of the Company.

4. Certificate of Ronald M. Simons, independent certified accountant for the Company as to changes in financial conditions of the Company for the period from December 31, 1984 to the Closing Date.

5. Letter by Michael Zinman forgiving the Company's debt to him in the amount of $61.00.

6. Letter of Resignation and Release of Michael Zinman as President, Secretary and sole Director of the Company.

7. Stock Certificate No. 1, representing one (1) share (no par value) of the common stock of the Company, duly endorsed in bearer form by Seller. A receipt therefor was issued by Buyer.

B. <u>Buyer delivered to Seller</u>:

1. Certified Check in the amount of U.S. $500.00 payable to the order of Michael Zinman. A receipt therefor was issued by Seller.

C. <u>The following were also delivered to Buyer and a receipt therefor issued by Buyer;</u>

1. Certificate of Incorporation of the Company together with Certificate of the Secretary of State of the State of New York dated on July 24, 1984.

2. By-Laws of the Company duly adopted at the organization meeting of the Company on February 13, 1984.

3. Minutes/Record Book of the Company.

4. Seal of the Company.

All actions necessary to effectuate the Closing have been taken, the Closing was deemed ended at 5:00 P.M., the 11th day of April, 1985.

-2-

CERTIFICATE OF MICHAEL ZINMAN

I, Michael Zinman, as Seller and as President of Raterpiller Tractor Co., Inc., a New York corporation, do hereby certify that as of the date hereof all the representations, warranties and covenants contained in Articles 4.1 through 4.31 of the Share Acquisition Agreement by and between Caterpillar Tractor Co., a California corporation, and myself, dated April 9, 1985, are true and correct, and have been duly complied with and fulfilled. No fact is known to me, nor do I have any reason to know of any fact, which would have the effect of making any of the above-mentioned representations and warranties misleading or which might reasonably be expected to have a material adverse effect on the Company.

IN WITNESS WHEREOF, I have hereunto set my hand this 11th day of April, 1985.

Michael Zinman

LAW OFFICES OF

DECHERT PRICE & RHOADS

477 MADISON AVENUE

NEW YORK, NY 10022

TELEX 84 5324 • BARDEP

(212) 308-4400

3400 CENTRE SQUARE WEST
1500 MARKET STREET
PHILADELPHIA, PA 19102
(215) 972-3400

38 SQUARE DE MEEUS, BTE 1
1040 BRUSSELS, BELGIUM
(02) 511 60 40

32 BEDFORD SQUARE
LONDON WC1B 3EX, ENGLAND
01. 631. 3383

1730 PENNSYLVANIA AVENUE, N.W.
WASHINGTON, DC 20006
(202) 783-0200

800 NORTH THIRD STREET
HARRISBURG, PA 17102
(717) 233-7947

April 11, 1985

Caterpillar Tractor Co.
Peoria, Illinois 61629

Acquisition of all the outstanding shares of Raterpillar Tractor Co., Inc. by Caterpillar Tractor Co.

Dear Sirs:

We have acted as counsel to Mr. Michael Zinman, the sole shareholder of Raterpillar Tractor Co., Inc. (the "Company"), in connection with the transactions contemplated by the Share Acquisition Agreement (the "Agreement") by and between you, Caterpillar Tractor Co., a California corporation, and Mr. Michael Zinman, dated April 9, 1985. Unless otherwise defined herein, capitalized terms used herein shall have the meanings assigned thereto in the Agreement. This opinion is rendered pursuant to Article 6.5 of the Agreement.

In connection with the opinions expressed below, we have examined an executed copy of the Agreement, copies of the Certificate of Incorporation and By-Laws of the Company, as certified to by the Secretary of the Company, as well as the Corporate Minute Book and Stock Transfer Ledger of the Company. We have also examined originals, or copies certified to our satisfaction, of such other documents, certificates and statements of government officials and corporate officers and other papers as we have deemed relevant or necessary as a basis for such opinions. In such examination, we have assumed the genuineness of all signatures and the authenticity of all documents submitted to us as originals and the conformity with the originals of all documents submitted to us as copies. As to any facts material to our opinions expressed herein, we have relied on the representations and warranties made in or pursuant to the Agreement and the other documents referred to therein, the accuracy of which we have not independently verified. In addition, we have, when relevant facts were not independently established by us, relied, to the extent we deemed such reliance proper, upon certificates of government officials and certificates and other written statements of officers of the Company.

Based on the foregoing and subject to the qualifications set forth below, it is our opinion that:

1. Seller has lawful authority to enter into the Agreement and to carry out the transactions contemplated thereby, including the sale and assignment of the Share to Buyer;

2. Neither the execution nor the performance of the Agreement by Seller has resulted or will result in a violation of or conflict with any term or provision of the Certificate of Incorporation or By-Laws of the Company or of any agreement or instrument known to us to which Seller or the Company is a party or by which Seller or the Company is bound. Such execution and performance will not violate or conflict with any other restriction, which is known to us or reasonably should be known to us, of any kind to which Seller or the Company is subject;

3. The Company is a corporation duly organized, validly existing and in good standing under the laws of the State of New York, and has full corporate power and lawful authority to carry on its business as now conducted and to own and operate its assets, properties and business;

4. The copies of the Certificate of Incorporation, attached to the Agreement as Exhibit A, and the By-Laws, attached to the Agreement as Exhibit B, of the Company, are true, complete and correct. The Certificate of Incorporation and By-Laws of the Company have not been changed or amended in any way during the period from the date of the Agreement through the Closing Date, and were duly adopted and filed with the Secretary of State of the State of New York;

5. Exhibit C to the Agreement is a true, complete and correct list of all the directors and officers of the Company, together with

Caterpillar Tractor Co. April 11, 1985
Page three

information as to any existing employment contracts or agreements;

6. The Company is qualified to do business in the State of New York, that jurisdiction being the only jurisdiction in which the Company has qualified to do business;

7. Seller is the sole holder of all of the issued capital stock of the Company. Seller's Share is validly issued, fully paid and non-assessible. To our knowledge, there are no existing options, calls or commitments of any character whatsoever, or agreements to grant the same, relating to shares of the Company's capital stock, nor are there outstanding any securities convertible into or exercisable for such shares; and

8. The Company does not own directly or indirectly any voting securities of any corporation or association. No company has been a subsidiary of the Company, and there is no company having any liability for which the Company is or will be legally responsible.

Very truly yours,

Dechert Price + Rhoads

Dechert Price & Rhoads

Apr 12,85 13:36 CT ALB

76s

P.01

State of New York
Department of State } ss:

I Hereby Certify, *that I have made diligent examination of the index of corporation papers filed in this Department for a certificate, order or record of a dissolution of*

RATERPILLAR TRACTOR CO., INC.

the certificate of incorporation of which corporation was filed January 26, 1984, with perpetual duration,

and that upon such examination, I find no such certificate, order or record, and that so far as indicated by the records of this Department, such corporation is a subsisting corporation.

Witness *my hand and the official seal of the Department of State at the City of Albany, this* 11th *day of* April *one thousand nine hundred and* eighty-five

Secretary of State

RONALD M. SIMONS
CERTIFIED PUBLIC ACCOUNTANT
9 DREXEL COURT
SPRING VALLEY, N. Y. 10977
(914)-352-6407

April 9, 1985

Caterpillar Tractor Co.
Peoria, Illinois 61629

Gentlemen:

I have reviewed the books and records of Raterpillar Tractor Co., Inc. for the period from January 1, 1985 through April 8, 1985 and have made inquiries of the principal officer of the Company and reviewed the minutes of the Company for that same period. Nothing has come to my attention as a result of the above that would have a material effect on the December 31, 1984 financial statements.

Please note that the tax liability of $293 that appears on the December 31, 1984 balance sheet has been paid personally by Michael Zinman.

April 11, 1985

Raterpillar Tractor Co., Inc.
Ardsley, New York

Gentlemen:

I, Michael Zinman, hereby release, acquit, remise and forever waive and discharge Raterpillar Tractor Co., Inc. (the "Company"), a New York corporation, from any and all claims, demands, actions, causes of action, damages and costs of every kind and nature in any way relating to my loan to you in the principal amount of Sixty-One Dollars (U.S. $61.00) as indicated on the Balance Sheet of the Company for the fiscal year ending December 31, 1984.

Michael Zinman

April 11, 1985

Gentlemen:

I hereby tender my resignation as President, Secretary and sole Director of Raterpillar Tractor Co., Inc., (the "Company"), said resignation to be effective immediately.

I hereby release, acquit, remise and forever waive and discharge the Company and Caterpillar Tractor Co., a California corporation, from any and all claims, demands, actions, causes of action, damages and costs of every kind and nature in any way relating to my employment in the above-mentioned offices of the Company.

Michael Zipman

INCORPORATED UNDER THE LAWS OF THE STATE OF NEW YORK

RATERPILLAR TRACTOR CO., INC.

The Corporation is authorized to issue 200 Common Shares without par value of one class

This Certifies that MICHAEL ZINMAN is the owner of one fully paid and non-assessable Shares of the above Corporation transferable only on the books of the Corporation by the holder hereof in person or by duly authorized Attorney upon surrender of this Certificate properly endorsed.

In Witness Whereof, the said Corporation has caused this Certificate to be signed by its duly authorized officers and to be sealed with the Seal of the Corporation.

Dated February 14, 1984

CATERPILLAR TRACTOR CO.

Peoria, Illinois 61629-7310

RECEIPT

I, R. Rennie Atterbury III, Associate Counsel of Caterpillar Tractor Co., a California corporation ("Caterpillar"), do hereby acknowledge receipt on the date hereof of Stock Certificate No. 1, representing one (1) share (no par value) of the common stock of Raterpillar Tractor Co., Inc., a New York corporation, duly endorsed, transferred and assigned in bearer form, and delivered to Caterpillar.

Dated: April 11, 1985

CATERPILLAR TRACTOR CO.

By R. Rennie Atterbury

rra5274245

RECEIPT

I, Michael Zinman, do hereby acknowledge receipt on the date hereof of a certified check in the amount of U.S. $500.00 payable to my order from Caterpillar Tractor Co. in payment for one (1) share (no par value) of the common stock of Raterpillar Tractor Co., Inc., a New York corporation.

Dated: April 11 , 1985

Michael Zinman

CATERPILLAR TRACTOR CO.

Peoria, Illinois 61629-7310

RECEIPT

I, R. Rennie Atterbury III, Associate Counsel of Caterpillar Tractor Co., a California corporation, do hereby acknowledge receipt on the date hereof from Michael Zinman, President of Raterpillar Tractor Co., Inc., a New York corporation (the "Company") of the following items:

1. Original Certificate of Incorporation of the Company as filed on January 26, 1984, together with the Certificate of the Secretary of State of the State of New York dated July 24, 1984;

2. Original By-Laws of the Company as duly adopted at the organization meeting of the Company on February 15, 1984;

3. Minutes/Record Book of the Company; and

4. Seal of the Company.

Dated: April 11, 1985

CATERPILLAR TRACTOR CO.

By R. Rennie Atterbury

rra5273830

The check. . .

Commercial National Bank of Peoria
PEORIA, ILLINOIS

70-4/711

Re: Acct. #002-0508 Caterpillar

DATE April 8, 1985

No. 262296

PAY COMMERCIAL NATIONAL BANK 500 DOLS 00 CTS $ - - - -500.00- - -

TO THE ORDER OF - - -Michael Zinman- - - - - - - - - - -

CASHIER'S CHECK

AUTHORIZED SIGNATURE

⑈262296⑈ ⑆071100049⑆ ⑈090⑉0036⑈

The press comments. . .

Business

Gannett Westchester Newspapers/Monday, April 22, 1985

A game of Cat and Rat

By Lisa Iervolino
Business Staff

A small company, conceived by an offhand remark and nurtured by a sense of humor until it took on truly comic proportions, has been sold to a Fortune 500 corporation that apparently takes even jokes seriously.

The company, Raterpillar Tractor Co., was sold by Michael Zinman of Ardsley for $500 to the mighty Caterpillar Tractor Co of Peoria, Ill., manufacturer of heavy construction equipment.

Why did Caterpillar buy Zinman's company? The assets of Raterpillar, which had not yet begun operation, included two boxes of stationery and two rats, one white and one gray. Zinman said lightheartedly that he had considered making rat coats, thinking they could become a fad among punk rockers and other trend-setters. He also considered Rat of the Week and Adopt a Rat campaigns.

Zinman can't talk about the sale without dissolving into laughter. He says he'd love to know why Caterpillar bought Raterpillar. "Maybe this is part of their expansion plans," Zinman speculated. "Maybe they're planning to launch a surprise foray into a new industry. You never know."

Caterpillar is not talking. The matter is "too trivial for comment," a spokesman said.

Zinman said Raterpillar was formed after he said in jest to his attorneys, "We ought to form the Raterpillar Tractor Co." — and they did. When he received the certification of business, "I thought it was hilarious. I wrote to the general counsel at Caterpillar. It was a private spoof," he said.

comment on it, from 1977 to 1983 he had been involved in a trademark lawsuit between Caterpillar and Earthworm Tractor Co., a heavy equipment company he owns.

In the letter to Caterpillar, he wrote, "I am somewhat unsure as to (Raterpillar's) ultimate aims, goals and aspirations, but do not fret, they will come to me. However, I assure you it will not be in the area of the manufacture, sale or brokerage of heavy

> 'You'd think that a company that just lost $70 million in the first quarter of 1985 would have better things to do.'

construction equipment or other Caterpillar endeavors."

The Raterpillar letterhead bears a striking resemblance to the Caterpillar logo, with one addition: "A modest disclaimer: Raterpillar Tractor Co. is in no way connected with Caterpillar Tractor Co. Caterpillar sells 'Cats.' Raterpillar Tractor Co. sells rats."

Caterpillar's response was to suggest he change the company's name, because of the similarity. Zinman retorted, "I take exception to (your) position that Caterpillar could make Raterpillar bite the poisoned cheese."

Last April, he suggested that Caterpillar buy his company for

To his surprise, they agreed.

Over the course of a year, Zinman wrote repeatedly to Caterpillar, gently pulling the company's leg with cat and rat puns. When he had not heard from Caterpillar officials for some time, he inquired, "Cat got your tongue?"

After sending Caterpillar officials some T-shirts bearing the Raterpillar logo, he received a letter in July from Eugene Goodale, Caterpillar supervising patent attorney of trademarks and copyrights. In the letter, Goodale said, "As you mentioned to me some years ago, many large companies have lost a sense of humor. The Raterpillar T-shirt reaction ranged from laughter to !#&*$!!."

He suggested, "After seven straight quarters of no profit, I would venture to say that your sales and profit picture may be better than Caterpillar's. Maybe you should pay us to take this company off your hands — no sense pouring good money down a rathole."

Several months passed. Zinman wrote to George Schaefer, the newly elected Caterpillar chairman. "I thought we had a deal," he wrote. Having read an article in Fortune magazine about a Caterpillar deal to supply equipment to a Latin American logger in exchange for a shipment of coffins, he made a suggestion to cover Raterpillar's upkeep. He proposed a Christmas gift, a three-inch desktop coffin, emblazoned with the Cat "C."

"We could even personalize the lid if you thought it not in too poor taste. The way things are going in Peoria, we could

Please see RATERPILLAR

C SECTION

Classified inside

Gannett Kathy Kmonicek

RATERPILLAR/ From page C1

probably...presell a lot of them to current or recently current employees. We could even do a special line for management."

Zinman recalled, "We were in a limited frame of lunacy, until the time came for the closing. I assumed they would send me a letter, and I would sign it and return it."

He said all he really wanted was to be able to put an ad in the newspaper showing a tombstone with the words: "Caterpillar Tractor Co. has acquired from Michael Zinman all of the outstanding shares of Raterpillar Tractor Co."

He did run the ad, but not until after a lengthy, formal closing procedure. A Caterpillar attorney flew to New York to close the deal, with a closing document that is 53 pages long. "As though this were a multimillion dollar purchase," Zinman said in disbelief.

Noting that Caterpillar had reported losses for the last three years, including a $428 million deficit in 1984, he said, "You'd think that a company that just lost $70 million in the first quarter of 1985 would have better things to do."

When the sale was completed, Zinman turned over the stationery and the rats, which were claimed by a secretary in the attorney's office after Caterpillar's attorney refused to bring them back to Peoria.

He says he is enjoying the sensation caused by the tongue-in-cheek press release he sent out announcing Caterpillar's new acquisition, and he is also enjoying reports of Caterpillar's discomfort regarding their unexpected publicity.

Frank Prezelski, a financial analyst at Shearson Lehman Brothers who follows Caterpillar, laughed when he heard about its acquisition of Raterpillar, but said he is not surprised by the way they handled it. "It's a very formally structured company. They take everything very seriously. The way their legal department works, you have to fill out all kinds of forms and then they send a brace of lawyers. They wouldn't recognize the joke."

Of Caterpillar's financial standing, Prezelski said, "They've taken the position that whatever Caterpillar does is right, and the rest of the world is out of step. But there's a general feeling around that Caterpillar has screwed up its strategic planning for the past 20 years. They have gone to great lengths to pin the blame on the nasty Japanese who have somehow conspired to keep the dollar high and the yen low. But the real problem is that the worldwide demand for construction equipment is very weak."

As much as Zinman has enjoyed running the Raterpillar Tractor Co., his primary business is the Earthworm Tractor Co., a 20-year-old company that buys and sells new and used construction equipment in the U.S. and overseas. Earthworm, which went public in December, reported sales of $27 million in 1984, up 22 percent from the previous year. Net income was $416,115, compared to a loss of $770,993 in 1983. Zinman predicts sales of $31 million for 1985. He plans to capitalize on market instability, as demand for construction equipment shifts from country to country, and on what he sees as Caterpillar's weakening market position.

Zinman is a stocky, balding man with a full, red beard and one eyebrow shooting perpetually skyward. He grinned widely when he said he didn't sit up nights hatching the Raterpillar saga. "Ideas like this just come to you. It's quite likely that it will never happen again."

"Life is nasty, brutish and short," he said, quoting Thomas Hobbes, a 16th-century philosopher. "If a certain amount of benign pleasure can be introduced, it certainly makes our times easier. Just the idea of the events transpiring offers a certain amount of enjoyment to a number of people."

L.A. Times, April 22, 1985

—In the 20 years Michael Zinman sold Caterpillar Tractor Co. equipment, the company showed no sign of a sense of humor, he said. So he created "Raterpillar Tractor Co." to tickle its funny bone. He didn't. "Caterpillar decided to view it in the peculiar way in which I think only Caterpillar can view things," Zinman, of Ardsley, N.Y., said. "I did not expect them to react the way they reacted." The company threatened to sue him for trademark infringement. Eventually the giant tractor company settled the matter by paying $500 to take over the smaller one. But it refused to absorb Raterpillar's assets: a couple of pet-store rats.

—CONNIE STEWART

The Hartford Courant, April 22, 1985

Michael Zinman, who had sold Caterpillar Tractor Co. equipment in Peoria, Ill., for 20 years, wanted to see if the company had a corporate sense of humor. So early last year he created "Raterpillar Tractor Co." to tickle its funnybone. It didn't. "Caterpillar decided to view it in the peculiar way in which I think only Caterpillar can view things," Zinman said. "I did not expect them to react the way they reacted," threatening a lawsuit for infringement of trademark. Eventually the matter was settled, with the giant tractor company paying $500 to take over the smaller one. Said Richard Stober, a spokesman for Caterpillar's trademarks division: "It's an insignificant, trivial matter. It deserves no further attention or time on our part and we are not going to make any further comment."

★

Reprinted by permission of *the Hartford Courant* April 22, 1985

Quad-City Times, Sunday, April 21, 1985

QUAD-CITY TIMES Sunday, April 21, 1985 5C

The Cat and the Rat

PEORIA, Ill. (AP) — In an age of hostile takeovers, this truly was a friendly game of corporate cat and mouse, uh, rat.

Caterpillar Tractor Co., in its most unusual diversification, paid $500 to acquire Raterpillar Tractor Co.

But the Peoria-based king of construction-equipment manufacturing did not want Raterpillar's only asset: a pair of rats from a pet shop.

A Raterpillar legal secretary got those.

Michael Zinman, a Caterpillar dealer for more than two decades in Ardsley, N.Y., said Friday he ventured into the uncharted waters of corporate mirth with Raterpillar because Caterpillar's "attitude and image is too dour."

Zinman said he used his whimsical company to tease Caterpillar in an effort to determine whether it had a corporate sense of humor.

He wrote Caterpillar to announce his new company, and promised there'd be no competition, saying: "Raterpillar sells rats. Caterpillar sells cats."

Did Caterpillar react with humor?

> 'Raterpillar sells rats. Caterpillar sells cats.'

"Caterpillar decided to view it in the peculiar way in which I think only Caterpillar can view things," Zinman said. "I did not expect them to react the way they reacted."

But did they show a sense of humor?

Caterpillar threatened to sue over what it called an infringement of Caterpillar's registered trademark name, he said.

That triggered a series of pun-filled letters back and forth between the Cat and the Rat.

"I take exception to Caterpillar's opinion that Caterpillar could make Raterpillar bite the poisoned cheese," one of his letters said, adding Caterpillar could buy Raterpillar for what Zinman had spent on it: $384.20.

That prompted a response from Eugene Goodale, supervising patent attorney for Caterpillar's trademarks and copyrights section: "Maybe you should pay us to take this company off your hands — no sense pouring good money down a rat hole."

Goodale offered $200 for Raterpillar.

"That was the only genuinely funny letter Cat has ever written," Zinman said. He accepted the offer, writing: "We should put this rat to bed."

Months passed.

Finally, Caterpillar sent Raterpillar a 28-page document to close the deal. It sought details of Raterpillar's assets, an audited financial statement and other information.

"It was unbelievable. I mean, Caterpillar could use this closing document to buy the solar system, it was so complete," Zinman said.

He said Caterpillar demanded several revisions of the document and agreed to a $500 purchase price to cover extra legal fees the revisions brought Zinman.

"I don't know how many lawyers they had working on this, but it must have been an army," he said.

Finally everything was worked out. The sale was consummated a week ago.

Is Caterpillar dour, as Zinman said?

"We are not going to comment on that," said Richard Stober, a spokesman for Caterpillar's trademarks division.

Stober said of the Raterpillar episode: "It's an insignificant, trivial matter. It deserves no further attention or time on our part and we are not going to make any further comment."

Nice rice — here's how

Don't rinse white rice before cooking it unless the package so advises. You'll wash off the enriching nutrients by doing this.

Journal Star, Peoria, April 19, 1985

Rats! Caterpillar pays joker

By JON MORGAN
of the Journal Star

Caterpillar Tractor Co., after a yearlong game of cat and mouse, has purchased a New York-based company whose sole assets were two rats.

The purchase of Raterpillar Tractor Co. was finalized Saturday when a Caterpillar lawyer signed the papers and handed over a check for $500. He declined the rats, though.

Raterpillar, a company with no products and no real plans for producing any, was formed in January of 1984 by accident.

"It really was formed not so much by mistake but by misunderstanding," said Michael Zinman, president of Raterpillar as of closing. "I never expected the company to be formed in my wildest dreams."

Zinman, chairman of an international trading firm in Ardsley, N.Y., which deals in construction equipment, said he made an "off hand" comment in late 1983 to his lawyer about forming a company called Raterpillar Tractor Co.

"It was an offhand remark that was taken more seriously than intended," he said. Months later, the lawyer returned with the certificate of incorporation showing the company registered at the address of his attorney's New York offices.

A check with the New York Secretary of State confirms the registration of Raterpillar on Jan. 26, 1984, with 200 shares of no-par value stock.

"What was I supposed to do with it? I took a look at the certificate and realized there was no way I was going to do business with it in a serious way because I was going to be sued by Caterpillar."

But he decided to have some fun, partly he said, because "I have always questioned the dour attitude of Caterpillar."

"I'm pleased to announce the formation of a new company..." he wrote Caterpillar's Peoria headquarters in March of 1984. Though Raterpillar's focus was unclear, he assured Caterpillar that it would not be making construction equipment.

He offered to run disclaimers in his ads saying "Raterpillar Tractor Co. is in no way connected with Caterpillar Tractor Co. Raterpillar sells rats. Caterpillar sells Cats."

But Caterpillar, a company whose markets have suffered more than nibbling by competitors, was not convinced. It said the name was a violation of its trademark and threatened legal action if it was not changed.

This set off a series of letters — some of them tongue-in-cheek — between Caterpillar and Raterpillar.

"I take exception to Caterpillar's opinion that Caterpillar could make Raterpillar bite the poisoned cheese," wrote Zinman. Not interested in changing the name of his venture, he offered to sell it for what he had spent in filing fees and other costs: $384.20.

"I would not dismiss a counter-offer out of hand," he added.

He calls the response he got "the only genuinely funny letter Cat has ever written."

An official in Caterpillar's trademark department wrote on July 9 that "after seven straight quarters of no profits, I would venture to say that your sale and profit picture may be better than Caterpillar's."

The official made a counter-offer of $200, which Zinman accepted with an Oct. 10 letter that said "We should put this rat to bed."

After several months of no reply,

Please see CAT, Page A3

Reprinted by permission of Jon Morgan and *The Peoria Journal Star*

CAT

Continued from Page A1

during which he was forced to pay New York state taxes of $293 on Raterpillar, he wrote George Schaefer, Caterpillar's newly-appointed chairman, in February.

He congratulated Schaefer on his new job and said he had read that Caterpillar was considering a barter with a Latin American customer to accept coffins for payment on some equipment.

Zinman suggested Raterpillar could be licensed to sell miniature coffins with the Caterpillar logo to be used as desk decorations.

"That got a response," he said. Caterpillar's legal department sent a 28 page closing document in which Zinman had to declare in detail the assets of the company and provide audited financial reports. The document was returned to his attorneys several times for revision.

Afterward, the company's lawyer told him they knew the document would drive up Zinman's legal bill.

Zinman brought two live rats to the closing, held in his New York lawyer's office — Raterpillar World Headquarters — but the Caterpillar representative said he would not take them back to Peoria. The rats were given to a secretary.

The taxes Zinman had to pay drove the price of Caterpillar's most unusual diversification to date from $200 to $500.

Not satisfied with that, Zinman took out an official-looking ad in Wednesday's New York Times announcing the closing. The Wall Street Journal refused a similar ad but suggested he send a press release to the news department.

Zinman, head of a 27-year-old international trading firm called Earthworm Tractor Co., declined to say how much the Raterpillar adventure cost him but he admits losing money on it.

But, he added, it was worth it. "There are really too few opportunities to have pure corporate pleasure.

"The real question is, why did Caterpillar have to bother? Why did they go to such extremes?"

Richard Stober, spokesman for Caterpillar, said "It's an insignificant, trivial matter. It deserves no further attention or time on our part and we are not going to make any further comment."

But an official of the company, who declined to be identified, noted that trademark violations are taken seriously by the company, though the matter was viewed with humor by some Caterpillar people.

"The purchase was made in lieu of litigation over a potential technical violation," said the official.

There was no word on who will lead the new subsidiary, though Zinman suggested in a press release that recently-retired chairman Lee Morgan be considered.

L'affaire Raterpillar had faded away. The summer came and settled in when lo! Bob Cohen sent me copies of letters he sent to Atterbury at Caterpillar on August 12th and 15th. It seems that Cat had been less than diligent in notifying the tax authorities of the corporate change.

More perplexing, however, was the letter Atterbury sent to Cohen on the 21st. It seems that Caterpillar is liquidating Raterpillar. If they do this, all my efforts will have been for naught, and the name once again will be in the public domain.

LAW OFFICES OF

DECHERT PRICE & RHOADS

477 MADISON AVENUE
NEW YORK, NY 10022
TELEX 84 8824 • BARDEP

(212) 308-4400

3400 CENTRE SQUARE WEST
1500 MARKET STREET
PHILADELPHIA, PA 19102
(215) 972-3400

38 SQUARE DE MEEUS, STE 1
1040 BRUSSELS, BELGIUM
(02) 511 60 40

52 BEDFORD SQUARE
LONDON WC1B 3EX, ENGLAND
01. 631. 3383

1730 PENNSYLVANIA AVENUE, N.W.
WASHINGTON, DC 20006
(202) 783-0280

900 NORTH THIRD STREET
HARRISBURG, PA 17102
(717) 233-7947

ROBERT A. COHEN

August 12, 1985

R. Rennie Atterbury, III, Esq.
Associate Counsel
Caterpillar Tractor Company
Peoria, Illinois 61629-7310

Re: Caterpillar Tractor Co. Acquisition of Raterpillar Tarctor Co.

Dear Mr. Atterbury:

Enclosed is a "Notice of Employer Identification Number Assigned" with regard to Raterpillar Tractor Company. You might want to notify the IRS of a change in the address of record for Raterpillar so that there will be no delay in communications from the service.

Sincerely,

Robert

Robert A. Cohen

RAC:mc
Encl.

3400 CENTRE SQUARE WEST
1500 MARKET STREET
PHILADELPHIA, PA 19102
(215) 972-3400

36 SQUARE DE MEEUS, STE 1
1040 BRUSSELS, BELGIUM
(02) 511 60 40

32 BEDFORD SQUARE
LONDON WC1B 3EX, ENGLAND
01. 631. 3383

LAW OFFICES OF
DECHERT PRICE & RHOADS
477 MADISON AVENUE
NEW YORK, NY 10022
TELEX 64 6324 • BARDEP
(212) 308-4400

1730 PENNSYLVANIA AVENUE, N.W.
WASHINGTON, DC 20006
(202) 783-6200

800 NORTH THIRD STREET
HARRISBURG, PA 17102
(717) 233-7947

ROBERT A. COHEN

August 15, 1985

R. Rennie Atterbury, III, Esq.
Associate Counsel
Caterpillar Tractor Company
Peoria, Illinois 61629-7310

Re: Raterpillar Tractor Co.

Dear Rennie:

Enclosed is your "Federal Tax Deposit Coupon Book" relating to Raterpillar. Again, it is probably time to have the address of record changed.

Best regards.

Sincerely,

Robert

Robert A. Cohen

RAC:mc
Encl.

cc: Mr. Michael Zinman

LAW OFFICES OF

DECHERT PRICE & RHOADS

477 MADISON AVENUE
NEW YORK, NY 10022
TELEX 64 5824 · BARDEP

(212) 308-4400

3400 CENTRE SQUARE WEST
1500 MARKET STREET
PHILADELPHIA, PA 19102
(215) 972-3400

36 SQUARE DE MEEUS, STE 1
1040 BRUSSELS, BELGIUM
(02) 511 60 40

52 BEDFORD SQUARE
LONDON WC1B 3EX, ENGLAND
01. 631. 3383

1730 PENNSYLVANIA AVENUE, N.W.
WASHINGTON, DC 20006
(202) 783-0200

800 NORTH THIRD STREET
HARRISBURG, PA 17102
(717) 233-7947

ROBERT A. COHEN

August 26, 1985

Mr. Michael Zinman
Earthworm Tractor Co. (Inc.)
495 Ashford Avenue
Ardsley, New York 10502

Re: <u>Raterpillar</u>

Dear Michael:

Enclosed is a copy of a letter to me from our friend Rennie.

Best regards.

Sincerely,

Robert A. Cohen

RAC:mc
Encl.

CATERPILLAR TRACTOR CO.

Peoria, Illinois 61629

August 21, 1985

Mr. Robert A. Cohen
Dechert Price & Rhoads
477 Madison Avenue
New York, N.Y. 10022

Dear Robert:

I returned from an extended business trip to find your letters of August 12 and 15. Upon receipt of a tax clearance certificate from New York state, we should complete the process of liquidating Raterpillar Tractor Co. We will, of course, notify the IRS and other appropriate authorities of Raterpillar's tragic demise.

Best regards,

Rennie

Associate Counsel

RRAtterbury III
Telephone: (309) 675-4477
lr

rra51331100

Can you see it now?

SON OF RATERPILLAR

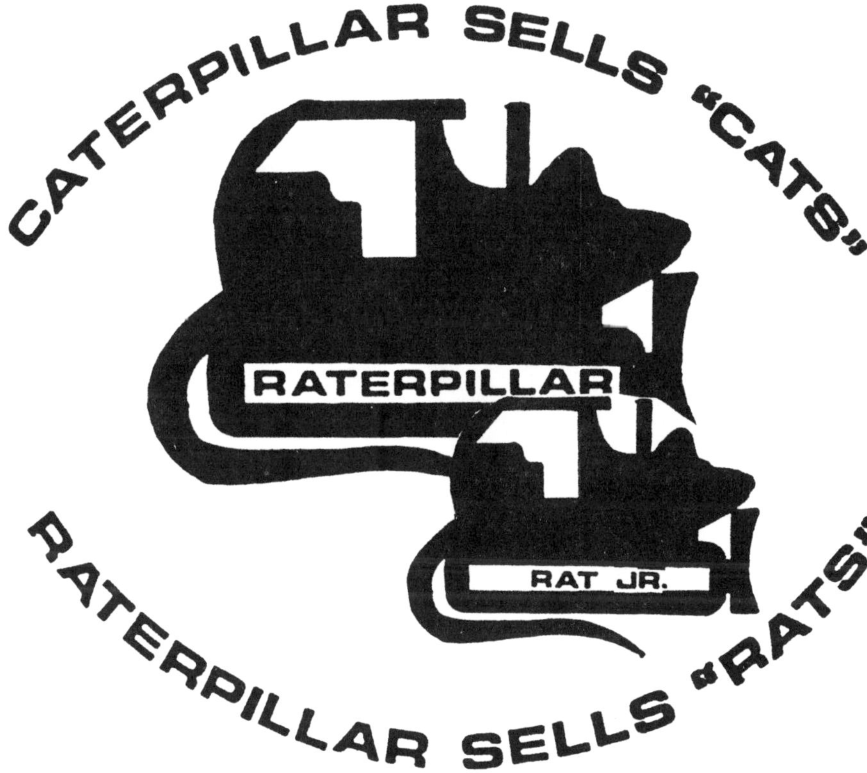

About the Author

Michael Zinman was president of the Earthworm Tractor Company for twenty years. A noted bibliophile, he is the author and editor of several books, among them The Massachusetts Laws of 1670, and American Color Plate Books by Whitman Bennett. His collection of the history of printing in the British Colonies and United States was the largest formed in the 20th century. It is now housed at the Library Company of Philadelphia, America's first public library. His collection of American Canvasser's books is now at the University of Pennsylvania library. The catalog of the collection, Canvassing Books in the Collection of Michael Zinman, is the definitive work on the genre. He now heads Japan, Inc., an energy company.